Unit IV

Phil Schlemmer

Illustrated by Patricia A. Sussman

LEARNING ON YOUR OWN!

Individual, Group, and Classroom
Research Projects for
Gifted and Motivated Students

The Center for Applied Research in Education, Inc.
West Nyack, New York

Library of Congress Cataloging-in-Publication Data

Schlemmer, Phillip L.
Mathematics projects.

(Learning on your own! : individual, group, and classroom research projects for gifted and motivated students ; unit 4)
1. Mathematics—Study and teaching. 2. Gifted children—Education—Mathematics. 3. Project method in teaching. 4. Independent study. I. Title. II. Series: Schlemmer, Phillip L. Learning on your own! ; unit 4.
QA11.S173 1987 372.7 86-24456

ISBN 0-87628-507-8

PRINTED IN THE UNITED STATES OF AMERICA

Dedication

This book is dedicated to my wife, Dori. Without her unending support, tireless editorial efforts, thoughtful criticisms, and patience, I could not have finished my work. Thank you, Dori.

Acknowledgments

My collaborator and co-teacher for eight years, Dennis Kretschman, deserves special mention at this juncture. Together we developed the activities, projects, and courses that became a "learning to learn" curriculum. Dennis designed and taught several of the projects described in these pages, and he added constantly to the spirit and excitement of an independent learning philosophy that gradually evolved into this set of five books. I deeply appreciate the contribution Dennis has made to my work.

I would also like to thank the following people for their advice, support, and advocacy: J. Q. Adams; Dr. Robert Barr; Robert Cole, Jr.; Mary Dalheim; Dr. John Feldhusen; David Humphrey; Bruce Ottenweller; Dr. William Parrett; Ed Saunders; Charles Whaley; and a special thanks to all the kids who have attended John Ball Zoo School since I started working on this project: 1973–1985.

About the Author

PHIL SCHLEMMER, M.Ed., has been creating and teaching independent learning projects since 1973, when he began his master's program in alternative education at Indiana University. Assigned to Grand Rapids, Michigan, for his internship, he helped develop a full-time school for 52 motivated sixth graders. The school was located at the city zoo and immediately became known as the "Zoo School." This program became an experimental site where he remained through the 1984–85 school year, with one year out as director of a high school independent study program.

Presently working as a private consultant, Mr. Schlemmer has been presenting in-services and workshops to teachers, parents, administrators, and students for more than 13 years and has published articles in *Phi Delta Kappan* and *Instructor.*

Foreword

This series of books will become invaluable aids in programs for motivated, gifted, and talented children. They provide clear guidelines and procedures for involving these children in significant learning experiences in research and high level thinking skills while not neglecting challenging learning within the respective basic disciplines of science, mathematics, social studies, and writing. The approach is one that engages the interests of children at a deep level. I have seen Phil Schlemmer at work teaching with the materials and methods presented in these books and have been highly impressed with the quality of learning which was taking place. While I recognized Phil is an excellent teacher, it nevertheless seemed clear that the method and the materials were making a strong and significant contribution to the children's learning.

Children will learn how to carry out research and will become independent lifelong learners through the skills acquired from the program of studies presented in these books. Success in independent study and research and effective use of libraries and other information resources are not simply products of trial-and-error activity in school. They are products of teacher guidance and stimulation along with instructional materials and methods and an overall system which provides the requisite skills and attitudes.

All of the material presented in this series of books has undergone extensive tryout. The author has also spent thousands of hours developing, writing, revising, and editing, but above all he has spent his time conceptualizing and designing a dynamic system for educating motivated, gifted, and talented youth. The net result is a program of studies which should make an invaluable contribution to the education of these youth. And, above all that, I am sure that if it is taught well, the kids will love it.

John F. Feldhusen, Ph.D., Director
Gifted Education Resource Institute
Purdue University
West Lafayette, Indiana 47907

About Learning on Your Own!

In the summer of 1973, I was offered the opportunity of a lifetime. The school board in Grand Rapids, Michigan authorized a full-time experimental program for 52 motivated sixth-grade children, and I was asked to help start it. The school was described as an environmental studies program, and its home was established in two doublewide house trailers that were connected and converted into classrooms. This building was placed in the parking lot of Grand Rapids' municipal zoo (John Ball Zoological Gardens). Naturally, the school came to be known as "The Zoo School."

The mandate for the Zoo School staff was clear—to build a challenging, stimulating, and interesting curriculum that was in no way limited by the school system's stated sixth-grade objectives. Operating with virtually no textbooks or "regular" instructional materials, we had the freedom to develop our own projects and courses, schedule our own activities, and design our own curriculum.

Over a period of ten years, hundreds of activities were created to use with motivated learners. This was a golden opportunity because few teachers are given a chance to experiment with curriculum in an isolated setting with the blessing of the school board. When a project worked, I wrote about it, recorded the procedures that were successful, filed the handouts, and organized the materials so that someone else could teach it. The accumulation of projects for motivated children led to a book proposal which, in turn, led to this five-book series. *Learning on Your Own!* is based entirely on actual classroom experience. Every project and activity has been used successfully with children in the areas of

- Research Skills
- Writing
- Science
- Mathematics
- Social Studies

As the books evolved and materialized over the years, it seemed that they would be useful to classroom teachers, especially in the upper elementary and junior high grades. This became increasingly clear as teachers from a wide variety of settings were presented with ideas from the books. Teachers saw different uses for the projects, based upon the abilities of their students and their own curricular needs.

Learning on Your Own! will be useful to you for any of the following reasons:

- If a curricular goal is to teach children to be independent learners, then skill development is necessary. The projects in each book are arranged according to the level of independence that is required—the early projects can be used to *teach* skills; the later ones require their *use.*
- These projects prepare the way for students confidently to make use of higher-level thinking skills.
- A broad range of students can benefit from projects that are skill-oriented. They need not be gifted/talented.
- On the other hand, teachers of the gifted/talented will see that the emphasis on independence and higher-level thinking makes the projects fit smoothly into their curricular goals.
- The projects are designed for use by one teacher with a class of up to 30 students. They are intentionally built to accommodate the "regular" classroom teacher. Projects that require 1-to-1 or even 1-to-15 teacher-student ratios are of little use to most teachers.
- The books do not represent a curriculum that must be followed. Gifted/talented programs may have curricula based upon the five-book series, and individual situations may allow for the development of a "learning to learn" curriculum. Generally speaking, however, each project is self-contained and need not be a part of a year-long progression of courses and projects.
- Each project offers a format that can be used even if the *content* is changed. You may, with some modification, apply many projects toward subject material that is already being taught. This provides a means of delivering the same message in a different way.
- Most teachers have students in their classes capable of pursuing projects that are beyond the scope of the class as a whole. These books can be used to provide special projects for such students so that they may learn on their own.
- One of the most pervasive concepts in *Learning on Your Own!* is termed "kids teaching kids." Because of the emphasis placed on students teaching one another, oral presentations are required for many projects. This reinforces the important idea that not only can students *learn,* they can also *teach.* Emphasis on oral presentation can be reduced if time constraints demand it.
- The premise of this series is that children, particularly those who are motivated to learn, need a base from which to expand their educational horizons. Specifically, this base consists of five important components of independent learning:

  —skills

  —confidence

  —a mandate to pursue independence

—projects that show students *how* to learn on their own

—an opportunity to practice independent learning

Learning on Your Own! places primary emphasis on the motivated learner, the definition of which is left intentionally ambiguous. It is meant to include most normal children who have natural curiosities and who understand the need for a good education. Motivated children are important people who deserve recognition for their ability and desire to achieve. The trend toward understanding the special needs and incredible potential of children who enjoy the adventure and challenge of learning is encouraging. Teachers, parents, business people, community leaders, and concerned citizens are beginning to seriously ask, "What can we do for these young people who want to learn?"

Creating a special program or developing a new curriculum is not necessarily the answer. Many of the needs of these children can be met in the regular classroom by teaching basic independent learning skills. No teacher can possibly master and teach all of the areas that his or her students may be interested in studying, but every teacher has opportunities to place emphasis on basic learning skills. A surprising number of children become more motivated as they gain skills that allow them to learn independently. "Learning on your own" is an important concept because it, in itself, provides motivation. You can contribute to your students' motivation by emphasizing self-confidence and skill development. One simple project during a semester can give students insight into the usefulness of independent learning. One lesson that emphasizes a skill can bring students a step closer to choosing topics, finding information, planning projects, and making final presentations without assistance. By teaching motivated students *how to learn on their own,* you give them the ability to challenge themselves, to transcend the six-hour school day.

Beyond meeting the immediate needs of individual students, teaching children how to learn on their own will have an impact on their adult lives and may affect society itself. It is easy to discuss the day-to-day importance of independent learning in one breath, and in the next be talking of the needs of adults 30 years from now. This five-book series is based upon the assumption that educating children to be independent learners makes sense in a complicated, rapidly changing, unpredictable world. Preparing today's children for tomorrow's challenges is of paramount importance to educators and parents, but a monumental task lies in deciding what can be taught that will have lasting value in years to come. What will people need to know in the year 2001 and beyond? Can we accurately prescribe a set of facts and information that will be *necessary* to an average citizen 10, 20, or 30 years from now? Can we feel confident that what we teach will be useful, or even relevant, by the time our students become adults? Teaching children to be independent learners is a compelling response to these difficult, thought-provoking questions.

How to Use Learning on Your Own!

Learning on Your Own! can be used in many ways. The projects and the overall design of the books lend themselves to a variety of applications, such as basic skill activities, full-class units or courses, small-group projects, independent study, and even curriculum development. Regardless of how the series is to be implemented, it is important to understand its organization and recognize what it provides. Like a good cookbook, this series supplies more than a list of ingredients. It offers suggestions, advice, and hints; provides organization and structure; and gives time lines, handouts, and materials lists. In other words, it supplies everything necessary for you to conduct the projects.

These books were produced with you in mind. Every project is divided into three general sections to provide uniformity throughout the series and to give each component a standard placement in the material. The first section, Teacher Preview, gives a brief overview of the scope and focus of the project. The second section, Lesson Plans and Notes, outlines a detailed, hour-by-hour description. After reading this, every nuance of the project should be understood. The third section, Instructional Materials, supplies the "nuts-and-bolts" of the project—reproducible assignment sheets, instructional handouts, tests, answer sheets, and evaluations.

Here is a concise explanation of each of the three sections. Read this material before going further to better understand how the projects can be used.

Teacher Preview

The Teacher Preview is a quick explanation of what a project accomplishes or teaches. It is divided into seven areas, each of which provides specific information about the project:

Length of Project: The length of each project is given in classroom hours. It does not take into account homework or teacher-preparation time.

Level of Independence: Each project is identified as "basic," "intermediate," or "advanced" in terms of how much independence is required of students. The level of independence is based primarily on how many decisions a student must make and how much responsibility is required. It is suggested that students who have not acquired independent learning skills, regardless of their grade level, be carefully introduced to advanced projects.

For teachers who are interested, there is a correlation between the skill development mentioned here and the progression to higher-level thinking skills typified by Benjamin Bloom's "Taxonomy of Educational Objectives":

Level of Independence	*Bloom's Taxonomy*
Basic	Knowledge Comprehension
Intermediate	Application Analysis
Advanced	Synthesis Evaluation

Goals: These are straightforward statements of what a project is designed to accomplish. Goals that recur throughout the series deal with skill development, independent learning, and "kids teaching kids."

During This Project Students Will: This is a list of concise project objectives. Occasionally, some of these statements become activities rather than objectives, but they are included because they help specify what students will do during the course of a project.

Skills: Each project emphasizes a specific set of skills, which are listed in this section. Further information about the skills is provided in the "Skills Chart." You may change the skill emphasis of a project according to curricular demands or the needs of the students.

Handouts Provided: The handouts provided with a project are listed by name. This includes assignment sheets, informational handouts, tests, and evaluation forms.

Project Calendar: This is a chart that graphically shows each hour of instruction. Since it does not necessarily represent consecutive days, lines are provided for you to pencil in dates. The calendar offers a synopsis of each hour's activity and also brief notes to clue you about things that must be done:

PREPARATION REQUIRED	STUDENTS TURN IN WORK
NEED SPECIAL MATERIALS	RETURN STUDENT WORK
HANDOUT PROVIDED	ANSWER SHEET PROVIDED

Lesson Plans and Notes

The lesson plan is a detailed hour-by-hour description of a project, explaining its organization and presentation methods. Projects can be shortened by reducing the time spent on such things as topic selection, research, and presentation; however, this necessitates de-emphasizing skills that make real independent study possible. Alternately, a project may require additional hours if students are weak in particular skill areas or if certain concepts are not thoroughly understood.

Each hour's lesson plan is accompanied by notes about the project. Some notes are fairly extensive if they are needed to clarify subject matter or describe a process.

Instructional Material

There are four types of reproducible instructional materials included in *Learning on Your Own!* Most projects can be run successfully with just a Student Assignment Sheet; the rest of the materials are to be used as aids at your discretion.

Student Assignment Sheets: Virtually every project has an assignment sheet that explains the project and outlines requirements.

Additional Handouts: Some projects offer other handouts to supply basic information or provide a place to record answers or research data.

Tests and Quizzes: Tests and quizzes are included with projects that present specific content. Since most projects are individualized, the activities themselves are designed to test student comprehension and skill development.

Forms, Charts, Lists: These aids are provided throughout the series. They are designed for specific situations in individual projects.

OTHER FEATURES OF LEARNING ON YOUR OWN!

In addition to the projects, each book in the series offers several other useful features:

Grade Level: A grade level notation of upper elementary, junior high, and/or high is shown next to each project in the table of contents. Because this series was developed with gifted/talented/motivated sixth graders, junior high is the logical grade level for most projects; thus, generally speaking, these projects are most appropriate for students in grades 6–8.

Skills Chart: This is a chart listing specific independent learning skills that may be applied to each project. It is fully explained in its introductory material.

General Notes

Examine the *structure* of the projects in each book, even if the titles do not fit specific needs. Many projects are so skill-oriented that content can be drastically altered without affecting the goals.

Many projects are dependent upon resource materials; the more sources of information, the better. Some way of providing materials for the classroom are to

- Ask parents for donations of books and magazines.
- Advertise for specific materials in the classified section of the newspaper.
- Check out library materials for a mini-library in the classroom.

- Gradually purchase useful materials as they are discovered.
- Take trips to public libraries and make use of school libraries.

Students may not initially recognize the value of using notecards. They will soon learn, however, that the cards allow data to be recorded as individual facts that can be arranged and rearranged at will.

"Listening" is included as an important skill in most projects. In lecture situations, class discussions, and when students are giving presentations, you should require students to listen and respect the right of others to listen.

Provide time for grading and returning materials to students during the course of a project. The Project Calendar is convenient for planning a schedule.

A visual display is often a requirement for projects in this series. Students usually choose to make a poster, but there are other possibilities:

mural	collage	demonstration	dramatization
mobile	model	display or exhibit	book, magazine, or pamphlet
diorama	puppet show	slide show	

When students work on their own, your role changes from information supplier to learning facilitator. It is also important to help students solve their own problems so that momentum and forward progress are maintained.

A FINAL NOTE FROM THE AUTHOR

Learning on Your Own! provides the materials and the structure that are necessary for individualized learning. The only missing elements are the enthusiasm, vitality, and creative energy that are needed to ignite a group of students and set them diligently to work on projects that require concentration and perseverance. I hope that *my* work will make *your* work easier by letting you put your efforts into quality and innovation. The ability to learn independently is perhaps the greatest gift that can be conferred upon students. Give it with the knowledge that it is valuable beyond price, uniquely suited to each individual, and good for a lifetime!

Phil Schlemmer

About This Book

This book emphasizes the use of basic mathematic and problem-solving skills; the projects are designed to supplement the regular mathematics curriculum. They were field tested with motivated sixth-grade students, and should have their greatest application at the upper elementary and junior high school levels. Most of the projects require that students work individually (or in small groups) to solve problems and achieve stated objectives. These activities add fun or challenge to the necessary process of calculating answers to basic arithmetic problems. It is important to show students that mathematics is a tool, and that the study of mathematics need not be confined to solving columns of problems from a textbook: the tool can be put to use, and the results can be satisfying and rewarding.

As you study the contents of this book, notice that each project offers *many* mathematics problems to be solved. In this respect, there is little difference between it and a regular mathematics textbook. The bottom line remains intact: students learn to compute by computing. The difference in this book is that each set of problems comprises a new challenge or goal. In the first project, Fractions and Decimals, students add, subtract, multiply, and divide fractions and decimals to calculate the length of line segments and the perimeters and areas of closed figures. The *goal* is to accumulate points by calculating answers swiftly and accurately. Structured like a game, this project makes fraction and decimal problem solving more palatable.

The remaining projects in the book also offer unique challenges: Simple Equations emphasizes self-improvement; Mathematics Treasure Hunt is a game that requires students to use a wide range of basic skills to calculate the length of a map trail that leads to Treasure Mountain; Sections and Acres illustrates how land is commonly divided, measured, and sold in the United States; Protractor-Compass-Ruler makes use of geometry and point location to create drawings; Tree Mapping allows students to gather data and make accurate maps; Land Surveying teaches students about angles, lines, distances, field notes, and accuracy as they conduct surveys; Mathematics in the Park gives students an opportunity to go outside to collect data as they take measurements and calculate the amount of wood in various trees in the park or playground; Independent/ Individualized Projects allow students to work on their own to fulfill the requirements of specially designed projects titled Opinion Poll, Creating Story Problems, and Mathematics from Current Events. Each of these projects covers skills and material that are emphasized at the upper elementary and junior high levels.

It may appear that such projects as Tree Mapping, Land Surveying, Sections and Acres, and Mathematics in the Park are too complicated or involved to be conveniently used. Do not draw conclusions until you have studied the projects. They are designed to be simple *or* complex, depending on how they are presented. Tree Mapping, for example, can be conducted on a piece of paper at a desk, on the floor in a classroom, in a gymnasium, or outside in the woods. The plans for a simple homemade surveyor's transit are included if you decide to do these projects on a large scale outside. (Building the transit is a mathematics project all in itself.) Be sure to read the Lesson Plans and Notes that are provided with each project. They will explain ways of simplifying projects.

Individualized learning in mathematics places special requirements on the teacher. Briefly, here is a list of suggestions and advice that will help ensure a positive experience with the projects that are described in this book:

1. Become familiar with the project before presenting it to students. Be sure it accomplishes what *you* want it to. If it is to be changed, do so carefully so that it still works with revised objectives. Carefully review the hour-by-hour lesson plans and notes that are provided and make a list of the materials needed to conduct each project. Most of the necessary student handouts and answer sheets are also provided: check them carefully so that you know their content.

2. Establish a set of behavior and procedure rules that specify how things will operate in the classroom, or outside, during project time. The rules must be fair, and once they are established, they must be upheld. Students should be informed that following directions is as much a skill as measuring angles. Rules are established so that projects can be completed. Those who break the rules are exhibiting self-discipline skill deficiencies.

3. Give yourself plenty of time for grading and returning materials to students during the course of a project. Study the hour-by-hour description that is provided with each project, and understand that these hours do not necessarily represent consecutive days of instruction.
4. Some of the projects require students to produce visual displays such as maps or graphs. Emphasis should be placed on accuracy, neatness, and quality.
5. Establish a policy concerning calculators: either everyone should have access to one, or else no one should be allowed to use one.
6. Many of the projects in this book require that each student, or small group, be equipped with one or more of the following items:
   a. protractor
   b. ruler
   c. compass
   d. drawing paper
   e. clipboard
   f. measuring tape
   g. clinometer (instructions provided)
   h. homemade transit (instructions provided)

Mathematics is a critical area of study in today's technological world. The greatest challenge for teachers is to show students that mathematics is useful and that it can be fun. In an age of calculators and computers one does not have to be a genius to solve complicated mathematics problems; the great difficulty lies in convincing students to go beyond solving problems that are given to them and to expand their use of mathematics to the real world. For this reason, most of the projects in this book are designed to put mathematics skills to use; to see them produce a result. Although research is minimized, these projects emphasize independent learning nonetheless: it is hoped that you will use this book to help students recognize their potential for using mathematics on their own.

THE SKILLS CHART

Mathematics Projects is based upon skill development. The projects are arranged according to the amount of independence required, and a list of skills is provided for every project in the book. A comprehensive Skills Chart is included here to help define the skill requirements of each project. Many of them are basic, common-sense skills that are already being taught in your classes.

The Skills Chart is divided into six general skill areas: research, writing and planning, problem solving, self-discipline, self-evaluation, and presentation. Reading is not included on the chart because it is assumed that reading skills will be used with virtually every project.

The key tells if a skill is prerequisite (#), primary (X), secondary (0), or optional (*) for each project in the book. These designations are based upon the way the projects were originally taught; you may want to shift the skill emphasis of a project to fit the needs of your particular group of students. It is entirely up to you to decide how to present a project and what skills to emphasize. The Skills Chart is only a guide.

Examination of the chart quickly shows which skills are important to a project and which ones may be of secondary value. A project may be changed or rearranged to redirect its skill requirements. The projects in this book are designed to *teach* the use of skills. If a project's Teacher Preview lists twenty skills, but you want to emphasize only three or four of them, that is a perfectly legitimate use of the project.

Evaluating students on their mastery of skills often involves subjective judgments; each student should be evaluated according to his or her *improvement* rather than by comparison with others.

A blank Skills Chart is included in the Appendix. This chart can be helpful in several ways:

- Students can chart their own skill progression through a year. Give them a chart and tell them to record the title of a project on the first line. Have them mark the skills *you* have decided to emphasize with the project. This way, students will see *exactly* what skills are being taught and which ones they are expected to know how to use. As projects are continued through the year, the charts will indicate skill development.

- Use the chart to organize the skill emphasis of projects that did not come from this book. Quite often, projects have the potential to teach skills, but they are not organized to do so. An entire course or even a curriculum can be organized according to the skill development on the chart.

- The Skills Chart can be used as a method of reporting to parents. By recording the projects and activities undertaken during a marking period in the left-hand column, a mark for each of the 49 skills can be given. For example, a number system can be used:

  1—excellent
  2—very good
  3—good
  4—fair
  5—poor

- A simpler method of reporting to parents is to give them a copy of the Skills Chart without marks and use it as the basis for a discussion about skill development.

Finally, most teachers have little or no experience teaching some of the skills listed on the chart. There is plenty of room for experimentation in the field of independent learning, and there are no established "correct" methods of teaching such concepts as problem solving, self-evaluation, and self-discipline. These are things that *can* be taught, but your own teaching style and philosophy will dictate how you choose to do it. The skills listed on this chart should be recognizable as skills that are worth teaching, even if you have not previously emphasized them.

SKILLS CHART: MATHEMATICS

Symbol	Key
#	*Prerequisite Skills* Students must have command of these skills.
X	*Primary Skills* Students will learn to use these skills; they are necessary to the project.
0	*Secondary Skills* These skills may play an important role in certain cases.
*	*Optional Skills* These skills may be emphasized but are not required.

	RESEARCH					WRITING & PLANNING			PROBLEM SOLVING											
	COLLECTING DATA	INTERVIEWING	LIBRARY SKILLS	LISTENING	OBSERVING	NEATNESS AND ORGANIZING	SETTING OBJECTIVES	SELECTING TOPICS	ACCURACY	BASIC MATHEMATICS SKILLS	DIVERGE-CONVERGE-EVALUATE	DRAWING STRAIGHT & PARALLEL LINES	FACTORING	FOLLOWING & CHANGING PLANS	IDENTIFYING PROBLEMS	LINEAR & ANGULAR MEASUREMENT	MEETING DEADLINES	MULTIPLICATION & ADDITION PROPERTIES	POINT LOCATION	SCALE DRAWING/MAPPING
FRACTIONS AND DECIMALS				X		X			X	#										
SIMPLE EQUATIONS				X		X			X	#			X					X		
MATHEMATICS TREASURE HUNT	X			X	X	X			X	#					X			#		
SECTIONS AND ACRES				X	X	X			X	#										
PROTRACTOR-COMPASS-RULER				X	X	X			X			X				X			X	0
TREE MAPPING	X			X	X	X			X							X			X	X
LAND SURVEYING	X			X	X	X			X			X				X			X	X
MATHEMATICS IN THE PARK	X			X	X	X			X	#					X					
OPINION POLL	X	X	*	X	X	X	X	X	X	#	X			X	X		X			
CREATING STORY PROBLEMS	X	#	X	X	X	X	X	X	X	#	X			X	#		X			
MATH FROM CURRENT EVENTS	X	#	X	X	X	X	X	X	X	#	X			X	#		X			

SKILLS CHART: MATHEMATICS

PROBLEM SOLVING								SELF-DISCIPLINE										SELF-EVALUATION				PRESENTATION						
SOLVING FOR AN UNKNOWN	USING A COMPASS	USING A HOMEMADE TRANSIT	USING A PROTRACTOR	USING A RULER/STRAIGHTEDGE	USING EQUATIONS AND RELATED SENTENCES	WORKING WITH FRACTIONS AND DECIMALS	WORKING WITH LIMITED RESOURCES	ACCEPTING RESPONSIBILITY	CONCENTRATION	CONTROLLING BEHAVIOR	FOLLOWING PROJECT OUTLINES	INDIVIDUALIZED STUDY HABITS	PERSISTENCE	SHARING SPACE	TAKING CARE OF MATERIALS	TIME MANAGEMENT	WORKING IN GROUPS	PERSONAL MOTIVATION	SELF-AWARENESS	SENSE OF "QUALITY"	SETTING PERSONAL GOALS	CREATIVE EXPRESSION	CREATING PRESENTATION STRATEGIES	DRAWING/SKETCHING/GRAPHING	POSTER MAKING	PUBLIC SPEAKING	SELF-CONFIDENCE	TEACHING OTHERS
						X			X	X	X	X	X				X	X	X	X							X	
X					X	X		X	X			X	X					X	X	X	X						X	
X					X	X		X	X	X	X	X	X				*	X		X							X	
X					X	X			X											X							X	
	X		X	X					X	X		X	X		X			X		X				X	0		X	
	X	*	X	X				X	X	X		X	X	X	X		X	X		X				X	*		X	
	X	*	X	X				X	X	X	X	X	X	X	X		X	X		X				X	*		X	
X					X		X	X	X	X	X	X	X		X	X	*	X		X							X	
X					X	X	X	#	X	#	#	#	X		#	#	*	#	X	X	X	X	X	X	X	X	X	X
X					X	X	X	#	X	#	#	#	X		#	#	*	#	X	X	X	X	X	X	X	X	X	X
X					X	X	X	#	X	#	#	#	X		#	#	*	#	X	X	X	X	X	X	X	X	X	X

Contents

M-1

FRACTIONS AND DECIMALS: A CLASSROOM GAME

Teacher Preview

General Explanation:

"Fractions" and "Decimals" are two separate projects, set up as games, that are identical in structure. Each requires four hours to complete. Students work individually, or in pairs, to solve the problems on five worksheets, one sheet at a time. Upon completing a worksheet, the student brings it to the teacher to be checked, and if all answers are correct the next worksheet is provided. A system is established that awards points for getting work done quickly and accurately on each worksheet. At the end of the game, students add their points and the greatest total "wins." The Teacher Preview and Lesson Plans and Notes provided here are designed to cover both games.

Length of Project: 4 hours

Level of Independence: Basic

Goals:

1. To practice basic multiplication and division skills.
2. To introduce students to three simple geometric terms: "perimeter," "area," and "line segment."
3. To allow students to work with fractions and decimals.

During This Project Students Will:

1. Solve problems that require multiplication and division of fractions or decimals.
2. Calculate the *perimeter* of closed geometric figures that are composed of straight lines.
3. Calculate the *areas* of rectangles.
4. Use correct *units* with the answers to all mathematical calculations that require them.

Skills:

Neatness	Self-awareness
Organizing	Sense of "quality"
Working in groups	Self-confidence

Concentration
Controlling behavior
Following project outlines
Individualized study habits
Persistence
Personal motivation
Multiplication of fractions/decimals
Division of fractions/decimals
Accuracy
Basic mathematics skills
Listening

Handouts Provided:

- "Student Answer-and-Point Sheet"
- "I. Fractions: Addition and Subtraction"
- "II. Fractions: Addition"
- "III. Fractions: Multiplication and Addition"
- "IV. Fractions: Multiplication"
- "V. Fractions: Division"
- "I. Decimals: Addition and Subtraction"
- "II. Decimals: Addition"
- "III. Decimals: Multiplication and Addition"
- "IV. Decimals: Multiplication"
- "V. Decimals: Division"

PROJECT CALENDAR:

HOUR 1: ____________ The game is explained. Students receive Answer-and-Point Sheets and the first worksheet. As worksheets are completed, new ones are provided. All materials are returned at the end of the hour. PREPARATION REQUIRED HANDOUTS PROVIDED ANSWER SHEET PROVIDED STUDENTS TURN IN WORK	**HOUR 2:** ____________ The game continues. HANDOUTS PROVIDED ANSWER SHEET PROVIDED STUDENTS TURN IN WORK	**HOUR 3:** ____________ The game continues. HANDOUTS PROVIDED ANSWER SHEET PROVIDED STUDENTS TURN IN WORK
HOUR 4: ____________ (Use if needed.) The game is concluded. HANDOUTS PROVIDED ANSWER SHEET PROVIDED STUDENTS TURN IN WORK	**HOUR 5:** ____________	**HOUR 6:** ____________
HOUR 7: ____________	**HOUR 8:** ____________	**HOUR 9:** ____________

Lesson Plans and Notes

HOUR 1: Explain the mathematics game, either for fractions or decimals, depending on which set of worksheets you plan to use. Give each student a "Student Answer-and-Point Sheet" and follow the rules outlined below to play the game. At the beginning of the game, give students the first worksheet: "Addition and Subtraction." At the end of the hour, collect the worksheets and answer sheets.

HOW THE GAME IS PLAYED

This game can be played with students paired as partners, or it can be played with students working individually. Here is how the game works:

1. Each pair (or individual) receives worksheet I. The two partners work together to calculate answers to the six problems. Both partners should do the calculations and then compare answers; partners check one another to ensure that each problem has been answered correctly. A "Student Answer-and-Point Sheet" is provided for students to record the answers and points they earn.
2. When both partners are confident that all six of their answers for worksheet I are correct, one of them brings the "Answer-and-Point" sheet to you to be checked. (You check only answers for this, not procedure or arithmetic.)
3. On your desk are five stacks of cards; each stack is labeled with a Roman numeral to correspond with one of the worksheets: I, II, III, IV, and V. The cards in each stack are numbered, one through fifty. These cards are placed upside down in their stacks; the bottom card of each stack (the card next to the desk) is "1" and the top card of each stack is "50."
4. Student-pairs can accumulate points by answering *all* of the problems on a worksheet correctly. For example, the first person to complete worksheet I comes to your desk: she hands you her Answer-and-Point sheet and her answers are all correct. She is told to take the top card from stack I and record the number (50) on her sheet. When this is done she returns her number card and is given worksheet II, which she and her partner can now begin working on. *When students return their number cards, put them aside so they won't become mixed up with the remaining cards in the stacks.*
5. The next person in line hands you his answer sheet and all of his answers are correct. He takes the next card from stack I and records the number (49) on his answer sheet. When he hands you the card, you give him worksheet II.
6. This continues until every pair has completed worksheet I and has continued to do as many worksheets as time allows. Each time an answer sheet is entirely correct (units and all!) the student takes the next card from the appropriate stack and records the points his or her team has earned.

7. At the end of the allotted game time (HOUR 4), have each pair add its points. Declare a winner—but emphasize that there are no losers!
8. At the end of each hour collect all materials so students can't work on problems at home.
9. Additional hints:
   a. It is very important to grade answers quickly so that calculation time isn't wasted standing in line. It is a good idea to have an aide, parent volunteer, or one or two students help grade and distribute worksheets.
   b. The point system is designed to reward both speed and accuracy. If an answer is incorrect, the pair must continue to work until they find the mistake and correct it. You may, if you wish, mark the answer that is incorrect, but it is more of a penalty for hurrying or carelessness if you simply hand it back and say "one of these answers is wrong." By not having all answers correct, the partners lose the next number on the stack, and when they return to have their answers checked a second time the number will be smaller because cards will have been taken. The trick to this game is to get the *correct* answers as quickly as possible.

HOURS 2, 3, and 4: Student materials are returned at the beginning of each hour, and students continue to play the game until they have completed all five problem worksheets. This may not require four hours.

General Notes About This Project:

- You can use the five worksheets for *each* game in a number of ways, in addition to the game described in Hour 1. Use them as in-class assignments or homework assignments. Or, give them to students who already have command of basic skills for adding, subtracting, multiplying, and dividing fractions, leaving extra time to work with slower learners. You may want to use the worksheets individually as quizzes or combine them into a test. They may also be useful as introductions to geometry (lines and shapes).
- The first project emphasizes fractions, so answers should be written as whole numbers and fractions, not as decimals. Fractions should always be reduced to lowest terms. The second set of worksheets covers decimals. The worksheets instruct students to round their answers for decimal problems to the nearest hundredth.
- Be sure to emphasize the use of units with the answers to problems on the worksheets. If an answer has no units, it is wrong.
- A teacher answer sheet is provided at the end of the student handouts. It gives the answers to the five "Fractions" worksheets and the answers to the five "Decimals" worksheets.

Name ______________________ Date ______________________

FRACTIONS OR DECIMALS GAME
Student Answer-and-Point Sheet

Your Partner's Name ______________________

This answer sheet is to be used for a mathematics game. The game may be played with partners or individually. If you have a partner, the two of you will be given a worksheet with several problems on it. *Working together,* solve all the problems; before handing in your answers, you should both agree that all of the answers are correct. After recording the answers on the Answer-and-Point Sheet, take it to your teacher to be checked.

1. If all of your answers are correct, you will be given a card with a number of points written on it. Record this number in the appropriate space below and return the card to your teacher. Upon returning the card, you will be given a second worksheet.
2. If there is one or more incorrect answer, the answer sheet will be returned to you without a point card. This means you and your partner must rework the problems to find your mistakes. After you have corrected the mistakes, bring your answer sheet back to the teacher to be checked. If they are all correct you will then receive a point card and record the number on your answer sheet.

Here are some tips for playing the game:

1. As more students have their answers checked, the point values of the cards get smaller. In other words, the first pair to get all the answers for worksheet I correct earns 50 points, the next pair earns 49 points, and so on.
2. Answers are wrong if they do not have units.
3. It is wise to have both partners do all of the problems, to check each other. It is better to take a little longer than to have to rework problems.

Name ______________________ Date ______________________

FRACTIONS OR DECIMALS GAME

Student Answer-and-Point Sheet (continued)

I. A = ______ D = ____ B = ______ E = ____ C = ______ F = ____	POINTS:
II. Greatest perimeter: Fig. __ = ____ Smallest perimeter: Fig. __ = ____	POINTS:
III. **Area** **Perimeter** A ______ ________ B ______ ________ C ______ ________ D ______ ________ E ______ ________	POINTS:
IV. **Figure** **Area** LARGEST _____ = ____ _____ = ____ _____ = ____ _____ = ____ SMALLEST _____ = ____	POINTS:
V. Unknown side for: A = ______ E = ____ B = ______ F = ____ C = ______ G = ____ D = ______ H = ____	POINTS:
	TOTAL POINTS: __________

Name ______________________________ Date ______________________

I. FRACTIONS: ADDITION AND SUBTRACTION

Each of the six *line segments* drawn below represents a length of $120\frac{1}{2}$ inches. Calculate how long each lettered segment is. All of your work should be done neatly on a separate piece of paper. Your answers should be recorded below and they *must be reduced to lowest terms*. Don't forget to include *units* with your answers!

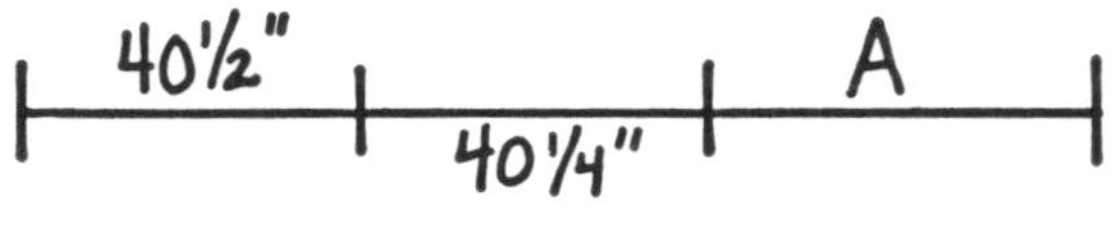

$40\frac{1}{2}$" | $40\frac{1}{4}$" | A

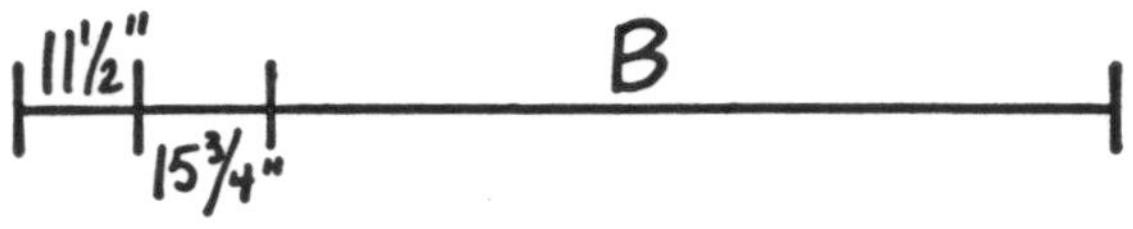

$11\frac{1}{2}$" | $15\frac{3}{4}$" | B

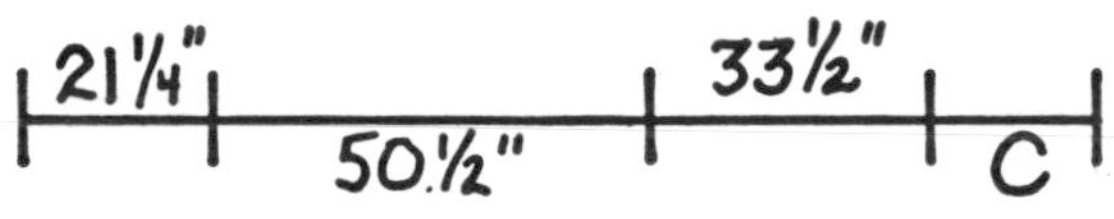

$21\frac{1}{4}$" | $50\frac{1}{2}$" | $33\frac{1}{2}$" | C

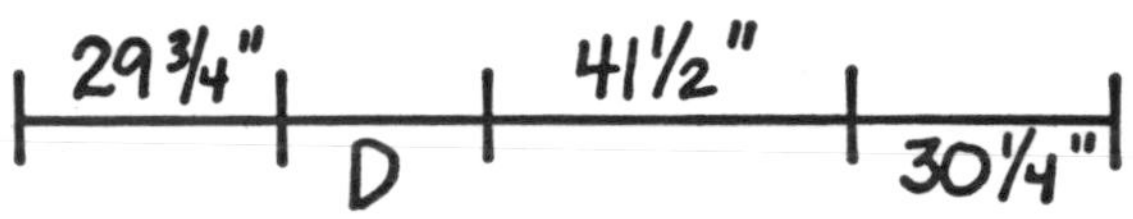

$29\frac{3}{4}$" | D | $41\frac{1}{2}$" | $30\frac{1}{4}$"

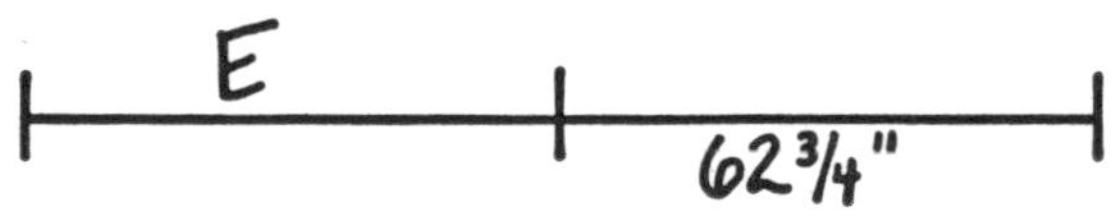

E | $62\frac{3}{4}$"

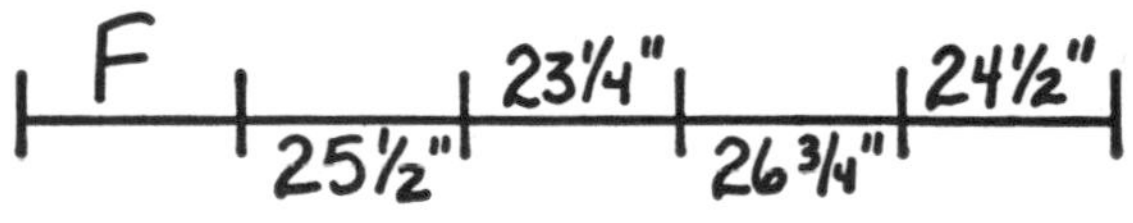

F | $25\frac{1}{2}$" | $23\frac{1}{4}$" | $26\frac{3}{4}$" | $24\frac{1}{2}$"

Segment Letter	Length
A	____________
B	____________
C	____________
D	____________
E	____________
F	____________

Name ______________________ Date ______________________

II. FRACTIONS: ADDITION

A "closed figure" is a drawing where all of the lines are connected but none of them cross, like the five examples below. The *perimeter* of a closed figure is the distance around its boundary. Calculate the perimeter of each figure and record which figure has the *greatest* perimeter and which figure has the *smallest* perimeter. All of your work should be done neatly on a separate piece of paper.

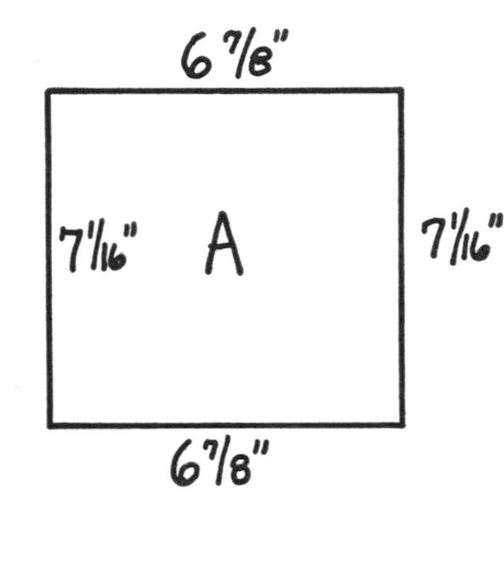

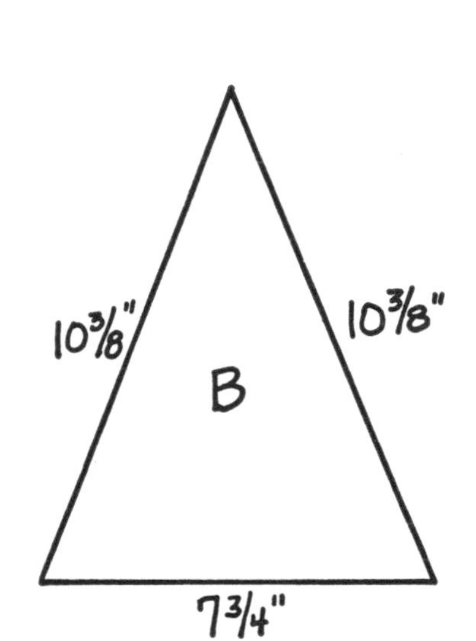

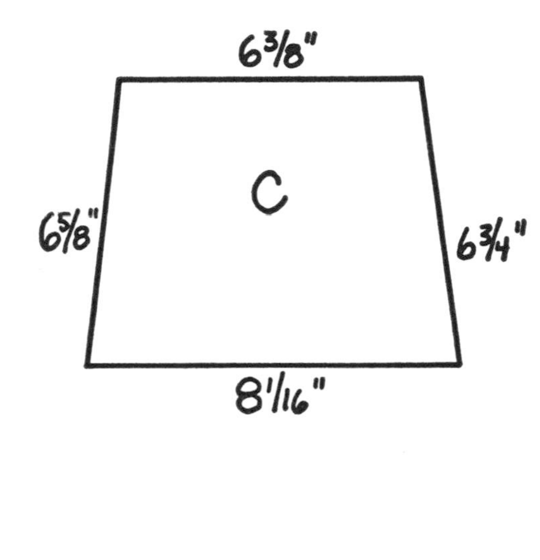

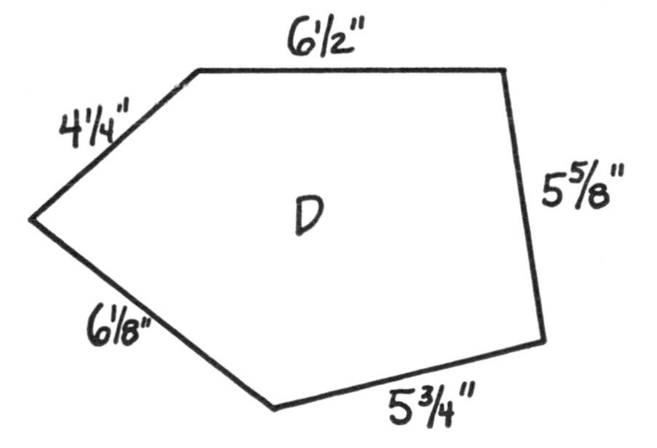

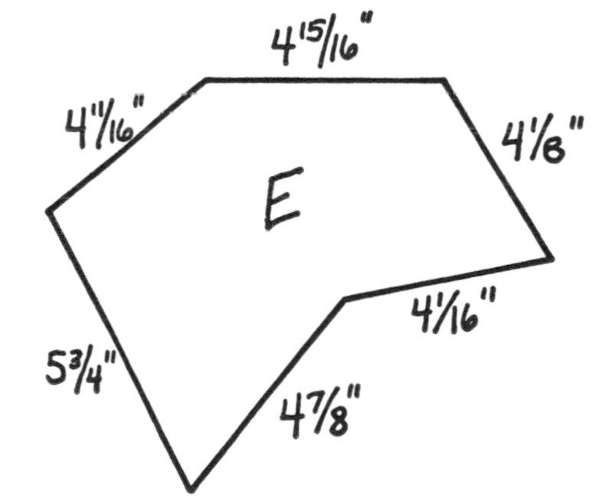

REMEMBER: Your answers must be reduced to lowest terms.

GREATEST PERIMETER: Figure ______________ = ______________________

SMALLEST PERIMETER: Figure ______________ = ______________________

Name ____________________ Date ____________________

III. FRACTIONS: MULTIPLICATION AND ADDITION

The figures below are all rectangles (a square is also a rectangle). To find the *area* of a rectangle, multiply the *length* times the *width* ($A = L \times W$). To find the *perimeter* of a rectangle, add each side ($P = L + W + L + W$). In the spaces below record the *area* and the *perimeter* of each rectangle. All of your work should be done neatly on a separate piece of paper. Reduce your answers to lowest terms. Since this project deals with units of inches, your answers for area must be in square inches ($in.^2$).

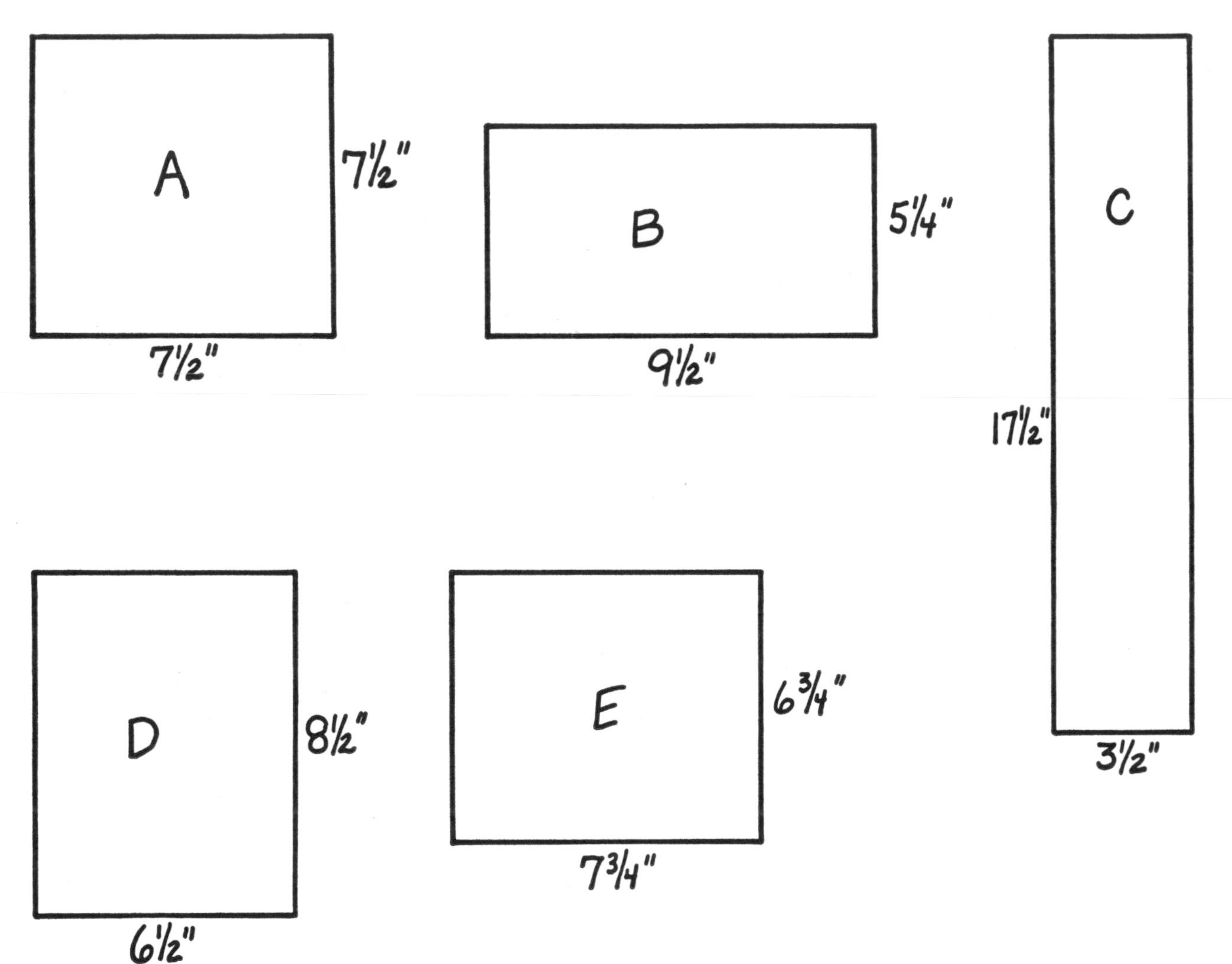

	Area	Perimeter
Figure A	____________	____________
Figure B	____________	____________
Figure C	____________	____________
Figure D	____________	____________
Figure E	____________	____________

Name ______________________ Date ______________________

IV. FRACTIONS: MULTIPLICATION

The figures below are all rectangles (a square is a rectangle). To find the *area* of a rectangle, multiply the length times the width (A = L × W). Calculate the *area* for the five rectangles below and rank them in size (area in square inches), from the largest to the smallest. All of your work should be done neatly on a separate piece of paper. Reduce your answers to lowest terms.

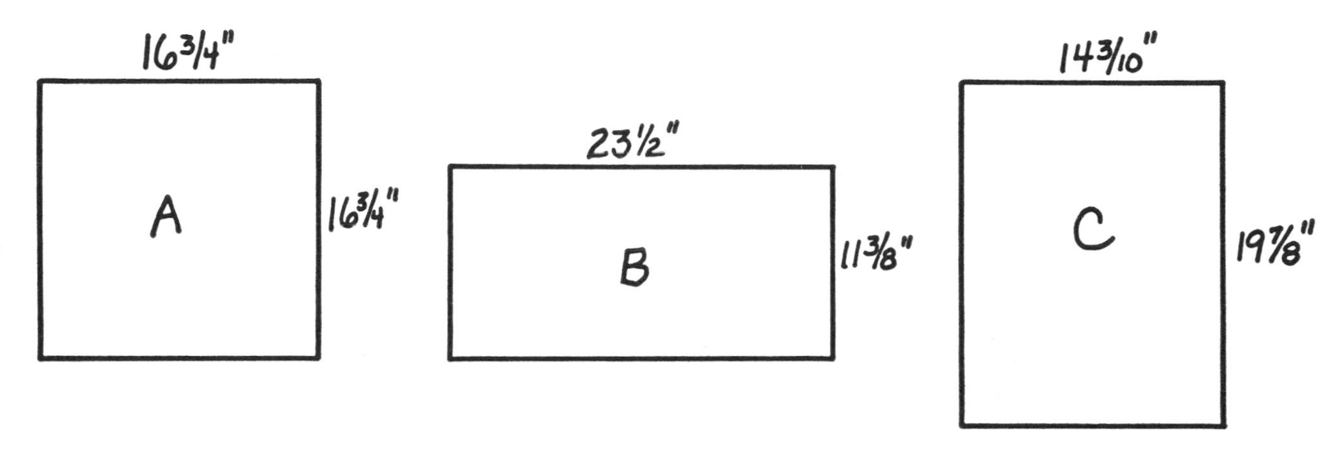

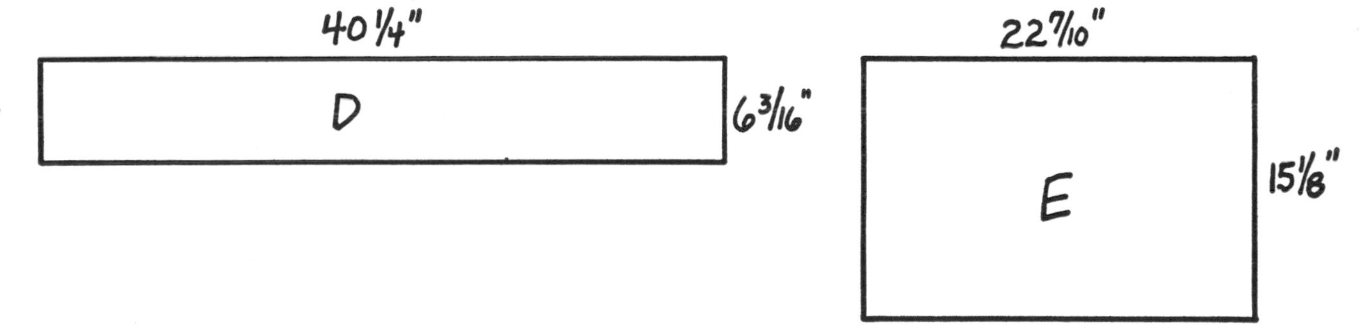

	Figure	Area
LARGEST:	________	________
	________	________
	________	________
	________	________
SMALLEST:	________	________

Name ______________________________ Date ______________________

V. FRACTIONS: DIVISION

The area of a rectangle can be calculated with this equation: $A = L \times W$, which can be rewritten in two ways: $L = A \div W$ and $W = A \div L$. If you know the area of a rectangle and how long one of the sides is, you can calculate how long the other side is by dividing. The figures below are all rectangles. Calculate how long either the length or the width is for each one. Record your answers in the space provided, along with the equation you used to calculate each answer. All of your work should be done neatly on a separate piece of paper. Remember to reduce your answers to lowest terms.

Figure	Equation	Answer
A	______________________	______________________
B	______________________	______________________
C	______________________	______________________
D	______________________	______________________
E	______________________	______________________
F	______________________	______________________
G	______________________	______________________
H	______________________	______________________

A: $6\frac{1}{2}$" (L); AREA = $32\frac{1}{2}$ IN.^{2}

B: AREA = $38\frac{1}{2}$ IN.^{2}; $5\frac{1}{2}$" (W)

C: $9\frac{1}{2}$" (L); AREA = $42\frac{3}{4}$ IN.^{2}

D: $8\frac{1}{2}$" (L); AREA = $72\frac{1}{4}$ IN.^{2}

E: $10\frac{1}{4}$" (W); AREA = $117\frac{7}{8}$ IN.^{2}

F: AREA = $120\frac{3}{8}$ IN.^{2}; $4\frac{1}{2}$" (W)

G: $8\frac{1}{4}$" (L); AREA = $63\frac{15}{16}$ IN.^{2}

H: AREA = $54\frac{27}{32}$ IN.^{2}; $6\frac{3}{4}$" (W)

Name ______________________________ Date ______________________

I. DECIMALS: ADDITION AND SUBTRACTION

Each of the six *line segments* drawn below is 150.50 cm long. Calculate how long each lettered segment is. All of your work should be done neatly on a separate piece of paper. Record your answers below and round them to the nearest hundredth if they have more than two decimal places. Don't forget to include *units* with your answers!

48.6 CM | 50.2 CM | A

11.1 CM | 15.8 CM | B

21.9 CM | 76.76 CM | 32.6 CM | C

38.21 CM | D | 47.57 CM | 37.45 CM

E | 77.639 CM

F | 28.19 CM | 31.68 CM | 33.586 CM | 27.876 CM

Segment Letter	Length
A	______________
B	______________
C	______________
D	______________
E	______________
F	______________

Name ______________________________ Date ______________________________

II. DECIMALS: ADDITION

A "closed figure" is a drawing where all of the lines are connected but none of them cross, like the five examples below. The *perimeter* of a closed figure is the distance around its boundary. Calculate the perimeter of each figure; record which figure has the *greatest* perimeter and which figure has the *smallest* perimeter. All of your work should be done neatly on a separate piece of paper. Round your answers to the nearest hundredth if they have more than two decimal places.

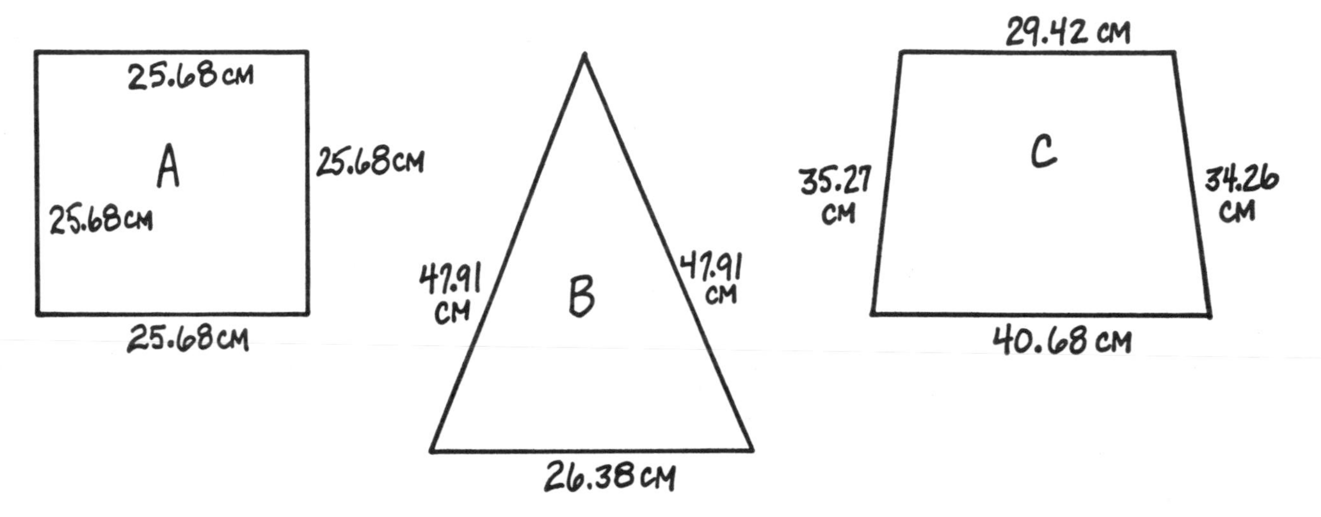

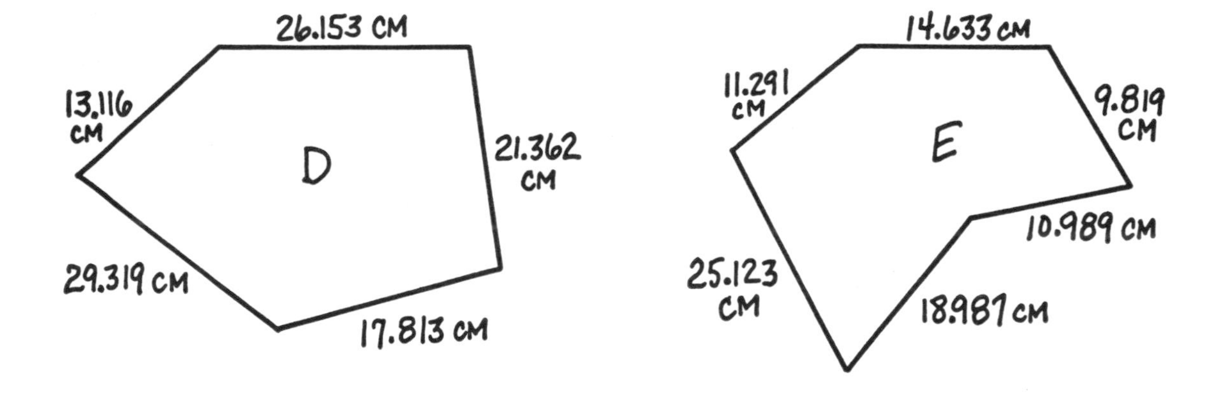

GREATEST PERIMETER: Figure ____________________ = ____________________

SMALLEST PERIMETER: Figure ____________________ = ____________________

Name ______________________ Date ______________________

III. DECIMALS: MULTIPLICATION AND ADDITION

The figures below are all rectangles (a square is a rectangle). To find the *area* of a rectangle, multiply the *length* times the *width* (A = L × W). To find the *perimeter* of a rectangle, add each side (P = L + W + L + W). In the spaces below record the *area* and the *perimeter* of each rectangle. All of your work should be done neatly on a separate piece of paper. Remember: your answers must be rounded to the nearest hundredth. Since this project deals with units of centimeters, your answers for area must be in square centimeters (cm^2).

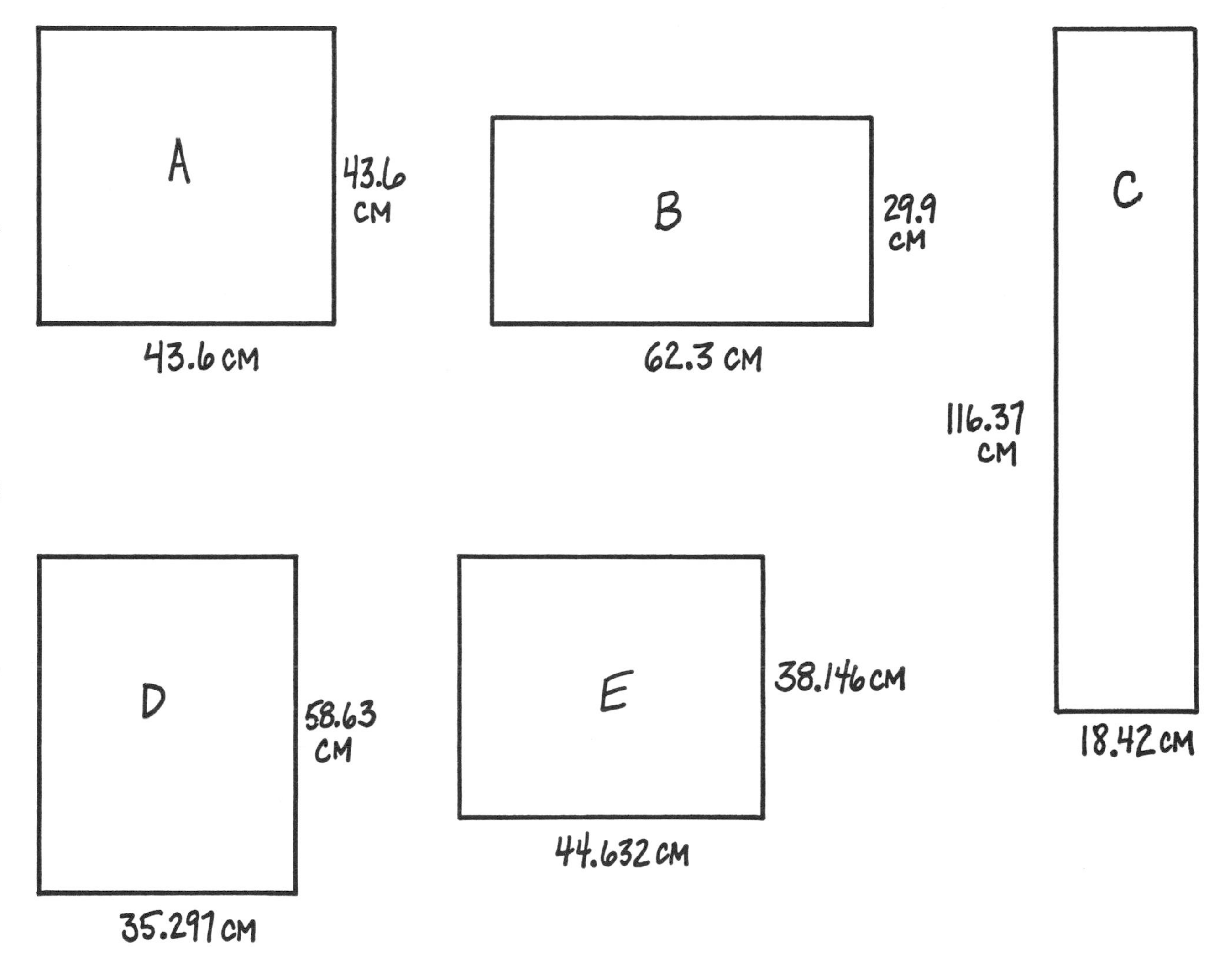

	Area	Perimeter
Figure A	________	________
Figure B	________	________
Figure C	________	________
Figure D	________	________
Figure E	________	________

Name ______________________ Date ______________________

IV. DECIMALS: MULTIPLICATION

The figures below are all rectangles (a square is a rectangle). To find the *area* of a rectangle, multiply the length times the width (A = L × W). Calculate the *area* for the five rectangles below and rank them by size (area in square centimeters), from the largest to the smallest. All of your work should be done neatly on a separate piece of paper. Remember to round your answers to the nearest hundredth.

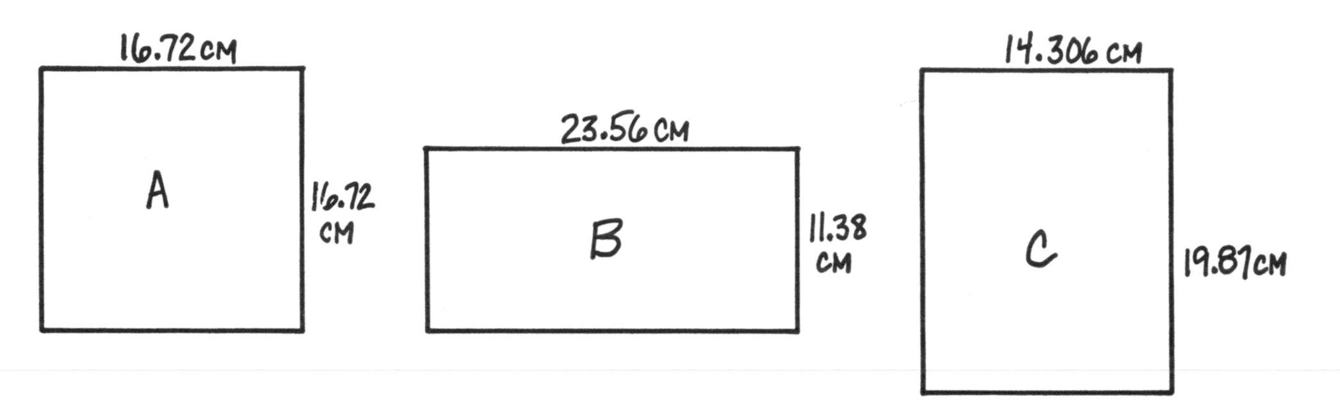

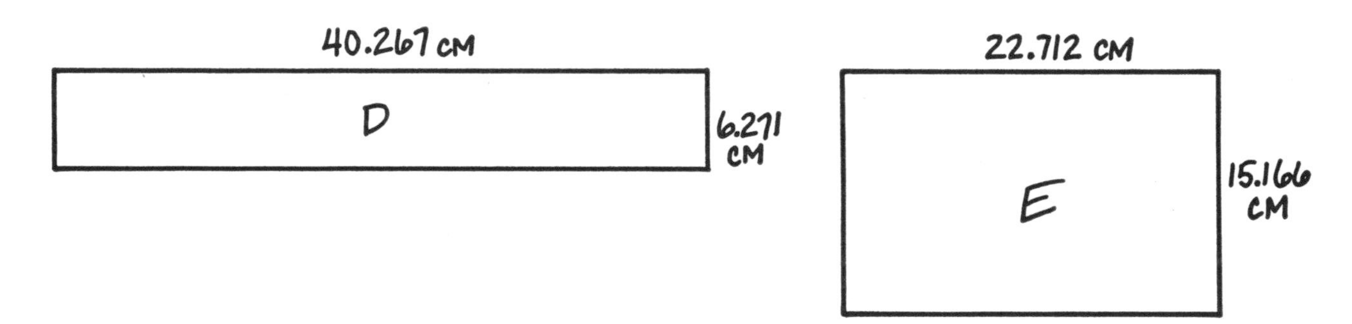

	Figure	Area
LARGEST:	______	______
	______	______
	______	______
	______	______
SMALLEST:	______	______

Name ______________________ Date ______________________

V. DECIMALS: DIVISION

The area of a rectangle can be calculated with this equation: $A = L \times W$, which can be rewritten in two ways: $L = A \div W$ and $W = A \div L$. If you know the area of a rectangle and how long one of the sides is, you can calculate how long the other side is by dividing. The figures below are all rectangles. Calculate how long either the length or the width is for each one. Record your answers in the space provided, along with the equation you used to calculate each answer. All of your work should be done neatly on a separate piece of paper. Remember to round your answers to the nearest hundredth.

Figure	Equation	Answer
A	______________________	______________________
B	______________________	______________________
C	______________________	______________________
D	______________________	______________________
E	______________________	______________________
F	______________________	______________________
G	______________________	______________________
H	______________________	______________________

6.5 CM (L)
A
AREA = 32.5 CM²

B
AREA = 38.6 CM²
5.8 CM (W)

9.37 CM (L)
C
AREA = 42.78 CM²

8.17 CM (L)
D
AREA = 72.23 CM²

10.72 CM (W)
E
AREA = 117.86 CM²

F
AREA = 120.83 CM²
4.16 CM (W)

8.306 CM (L)
G
AREA = 63.918 CM²

H
AREA = 54.248 CM²
6.764 CM (W)

FRACTIONS: TEACHER ANSWER SHEET

I. Addition and Subtraction (line segments)

A = $39\frac{3}{4}$ in.	D = 19 in.
B = $93\frac{1}{4}$ in.	E = $57\frac{3}{4}$ in.
C = $15\frac{1}{4}$ in.	F = $20\frac{1}{2}$ in.

II. Addition (perimeters)

A = $27\frac{7}{8}$ in.	D = $28\frac{1}{4}$ in.
B = $28\frac{1}{2}$ in.	E = $28\frac{7}{16}$ in.
C = $27\frac{13}{16}$ in.	

Greatest perimeter: B = $28\frac{1}{2}$ in.
Smallest perimeter: C = $27\frac{13}{16}$ in.

III. Multiplication and Addition (areas and perimeters)

	Area	Perimeter
A	$56\frac{1}{4}$ in.^{2}	30 in.
B	$49\frac{7}{8}$ in.^{2}	$29\frac{1}{2}$ in.
C	$61\frac{1}{4}$ in.^{2}	42 in.
D	$55\frac{1}{4}$ in.^{2}	30 in.
E	$52\frac{5}{16}$ in.^{2}	29 in.

IV. Multiplication (areas)

	Figure	Area
LARGEST	E	$343\frac{27}{80}$ in.^{2}
	C	$284\frac{17}{80}$ in.^{2}
	A	$280\frac{9}{16}$ in.^{2}
	B	$267\frac{5}{16}$ in.^{2}
SMALLEST	D	$249\frac{3}{64}$ in.^{2}

V. Division (areas)

Figure	Unknown Side
A	5 in.
B	7 in.
C	$4\frac{1}{2}$ in.
D	$8\frac{1}{2}$ in.
E	$11\frac{1}{2}$ in.
F	$26\frac{3}{4}$ in.
G	$7\frac{3}{4}$ in.
H	$8\frac{1}{8}$ in.

DECIMALS: TEACHER ANSWER SHEET

I. Addition and Subtraction (line segments)

A = 51.7 cm	D = 27.27 cm
B = 123.6 cm	E = 72.86 cm.
C = 19.24 cm	F = 29.17 cm

II. Addition (perimeters)

A = 102.72 cm	D = 107.76 cm
B = 122.20 cm	E = 90.84 cm
C = 139.63 cm	

Greatest perimeter: C = 139.63 cm
Smallest perimeter: E = 90.84 cm

III. Multiplication and Addition (areas and perimeters)

	Area	Perimeter
A	1900.96 cm^2	174.4 cm
B	1862.77 cm^2	184.4 cm
C	2143.54 cm^2	269.58 cm
D	2069.46 cm^2	187.85 cm
E	1702.53 cm^2	165.56 cm

IV. Multiplication (areas)

	Figure	Area
LARGEST	E	344.45 cm^2
	C	284.26 cm^2
	A	279.56 cm^2
	B	268.11 cm^2
SMALLEST	D	252.51 cm^2

V. Division (areas)

Figure	Unknown Side
A	5.0 cm
B	6.66 cm
C	4.57 cm
D	8.84 cm
E	10.99 cm
F	29.05 cm
G	7.70 cm
H	8.02 cm

M-2

SIMPLE EQUATIONS

Teacher Preview

General Explanation:

This is a "self-improvement" project. Students take a test to establish a base line for themselves on a self-improvement chart. Then they take two more tests to earn points which represent their degree of improvement on the chart.

Length of Project: 7 hours

Level of Independence: Basic

Goals:

1. To introduce students to the problem-solving process that is necessary to solve simple algebraic equations.
2. To place emphasis on basic mathematics skills.
3. To encourage students to strive for personal improvement.
4. To familiarize students with a number of common equations that they will encounter in later mathematics and science classes.

During This Project Students Will:

1. Rearrange equations and isolate unknown variables.
2. Solve simple equations that contain one unknown.
3. Establish a basic level of competence for solving equations.
4. Take at least two "Mathematics Self-Improvement" tests.

Skills:

Basic mathematics skills
Factoring
Addition properties
Multiplication properties
Solving for an unknown
Using equations
Listening
Neatness
Accepting responsibility
Concentration
Individualized study habits
Persistence
Personal motivation
Self-awareness
Sense of "quality"
Setting personal goals
Self-confidence
Accuracy
Working with fractions and decimals

Handouts Provided:

- "Equations Test"
- "Equations Self-Improvement Test 1"
- "Equations Self-Improvement Test 2"
- "Degree-of-Improvement Chart"
- "Equations Self-Improvement Test 3" (provides equations only, no values) You may use this to create your own test.

PROJECT CALENDAR:

HOUR 1: ______ Equations and problem solving are reviewed. Students are told they will take a timed test next hour.	**HOUR 2:** ______ Students take the Equations Test and record calculations and answers on separate paper. STUDENTS TURN IN WORK HANDOUT PROVIDED ANSWERS PROVIDED	**HOUR 3:** ______ Student work on the Equations Test is reviewed and discussed. Students are told that next hour they will take a second test. The goal is *improvement.* The Improvement Chart is discussed. RETURN STUDENT WORK HANDOUT PROVIDED
HOUR 4: ______ Students take Self-Improvement Test 1 and record calculations and answers on a separate paper. HANDOUT PROVIDED ANSWERS PROVIDED	**HOUR 5:** ______ Self-Improvement tests are discussed; each person's improvement over the Equations Test is marked on a chart. RETURN STUDENT WORK HANDOUT PROVIDED	**HOUR 6:** ______ Students take Self-Improvement Test 2 and record calculations and answers on a separate paper. HANDOUT PROVIDED ANSWERS PROVIDED
HOUR 7: ______ Improvement is marked on the chart, test material is returned to students, and the project is discussed. RETURN STUDENT WORK HANDOUT PROVIDED	**HOUR 8:** ______	**HOUR 9:** ______

Lesson Plans and Notes

HOUR 1: Give students a brief review of information they have already learned on rearranging equations and solving for an unknown variable. Describe the commutative and associative properties of multiplication and addition and explain the distributive property of multiplication. Work sample problems on the board to refresh students' memories about how to isolate a variable on one side of the equals sign. *It is assumed that students have already been taught these things.* At the end of the hour, tell the students that they will take a timed test on equations during the next hour; encourage them to study and come to class prepared to do their very best work.

Notes:

- It is assumed that students have been introduced to equations and problem solving before they begin this project. They should be able to solve for "n" in each of these instances (if values for the other letters are given):

$n = ab$	$a = \frac{n}{b}$
$n = \frac{a}{b}$	$a = \frac{b}{n}$
$n = a(b + c) + d$	$n + a = bc$
$a = nb$	$na = bc$

- This project may be used to *teach* students how to perform the operations needed to solve for "n" in the examples above. In this case, do the problems on the Equations Test in class, perhaps by having students do them on the board. Use the improvement tests in whatever way contributes most to your goals for the project.

HOUR 2: Students take the Equations Test and record their calculations and answers on notebook paper. Tell them to work carefully and steadily, but not to rush. Each problem must be "set up" properly (for two points) and answered correctly (for two more points). Emphasize neatness. At the end of the hour students hand in their papers with calculations and answers.

Notes:

- Since there are only 24 problems in each test, the total point value is 96 instead of 100. The extra 4 points should be awarded for neatness. It is important that students learn the value of neatness in mathematics, and this is one way to emphasize it.
- Encourage students to seriously concentrate on the Equations Test. Use whatever "gimmicks" you have up your sleeve to motivate them to work hard on this test. *Do not tell them that they will be taking future tests to measure self-*

improvement; they should think this test is all there is to the project, and that they are being evaluated on how well they do on it.

- It is important to point out to students that most people will not finish the Equations Test in the amount of time provided. Students should work methodically and carefully to get as many *correct* answers as possible.
- Require that each problem be solved step by step, and that each step be clearly shown. For example:

Equation:	$y = ax^2 + bx + c$	
Given:	$y = 58$	
	$a = 4$	
	$x = 3$	
	$c = 7$	
Find:	b	
	Solution:	$y = ax^2 + bx + c$
Step 1.	Substitute:	$58 = 4(3^2) + b(3) + 7$
Step 2.	Combine terms:	a) $58 = 4(9) + 3b + 7$
		b) $58 = 36 + 3b + 7$
		c) $58 = 43 + 3b$
Step 3.	Subtract 43 from each side:	$58 - 43 = 43 - 43 + 3b$
		$15 = 3b$
Step 4.	Divide each side by 3:	$\frac{15}{3} = \frac{3b}{3}$
		$5 = b$
Step 5.	Answer:	$b = 5$

You may not want your students' calculations to be this detailed, but they should show how they (1) substituted, (2) combined terms, (3) subtracted from each side of the equation, and (4) divided each side of the equation to find out what "b" equals.

HOUR 3: Spend this hour discussing the common mistakes found in grading the tests. You may want to let students keep their copy of the Equations Test for study purposes. Tell the students that next hour they will be given a self-improvement test. The equations and questions will be exactly like those on the Equations Test, but the numbers will be different. This is the first time that students learn about being evaluated on *improvement*. They are shown the "Degree-of-Improvement" chart and its function is explained. After the first test, everyone is at "0" on the chart. The next two tests will determine how much improvement can be achieved.

Note:

- This project is specifically designed to emphasize personal motivation and to reduce emphasis on peer competition. When the Equations Tests are graded, each student's score (no matter what it is) becomes his or her base line on a graph that is designed to show *degree of improvement*. Students are given the same test two more times (the same equations in the same order, but with different numbers so that answers cannot be memorized).

HOUR 4: Students are given the hour to take Self-Improvement Test 1.

HOUR 5: Discuss the tests and mark each person's improvement over the Equations Test on the chart.

Note:

- Each student's score on the first test becomes his or her "base" score, or "0," on the improvement chart. After this score is established, each subsequent test is graded in terms of improvement, and this *degree of improvement* is graphed on the chart. For example, if a student gets 20 points out of 100 on the first test and 55 points on the second test, he has improved his score by 35 points. He correctly completed 35% of the test that he did not complete on his initial attempt, so he is given 35 points on his improvement chart. If he gets 75 points on the third test, his increment of improvement is 20% more of the test, and he collects 20 more points on his chart.

HOUR 6: Students are given the hour to take Self-Improvement Test 2.

HOUR 7: Discuss the entire project and return the students' tests. Have each person mark his or her chart to indicate the degree of improvement over Self-Improvement Test 1. The finished chart represents how much each student was able to improve his or her initial score.

General Notes About This Project:

- Each problem on the tests is worth four points: two for the correct answer and two for the correct problem-solving procedure, or "setup." You determine how students should arrange their calculations on paper. Be sure to establish criteria that is specific enough to provide fair and understandable evaluation guidelines. All calculations and answers are recorded on separate pieces of paper. Remember that you must add 4 points for neatness to make this a 100-point test.
- Three separate tests are provided with this project. The equations are kept in the same order on each test and the same variable is unknown for the same problem on each test. The difficulty of the numbers (number of digits and decimal places) is consistent for all three tests. This should ensure that the scores reflect increasing ability to manipulate and solve equations.
- You may wish to increase the number of sessions in this project by having students take each test twice, or by creating new tests yourself. You may also wish to present the material differently; as homework material perhaps, or for mathematics races or team competitions. The tests are also useful simply as classroom worksheets.

- The equations in this project are all taken from physics and mathematics. To avoid confusion, there are no units in the problems; students should understand that units would be required in the answers to real-world problems.
- The Degree-of-Improvement Chart is provided to record students' improvement, but it is more effective to make a large poster-size chart to put on a wall in the classroom.
- Part of the value of this project is that it encourages self-improvement through self-motivated study. Be sure to emphasize the value of studying at home, and encourage students to seriously try to get better at solving the equations on the test.
- If students are allowed to use calculators, be sure that everyone has equal access to one.
- You may want to give students more than one hour to work on the tests. Let them work for one hour, collect the tests and calculation papers, and hand them out for a second hour of work during the next class period. You may also send them home, but in this case you are never quite certain who did the work.
- A third self-improvement test is included without values for the given variables. This may be used to create new test problems, emphasizing anything you want: more difficult numbers, rounding, solving for different variables, and so on.
- The following list shows each equation used in this project and explains what the letter variables stand for. This adds the potential for an entirely new project based upon actual data collection and problem solving, or perhaps an additional hour in which the equations are explained and their usefulness in the real world is described.

EQUATIONS USED IN "SIMPLE EQUATIONS"

1. $A = L \times W$	A = area of a rectangle L = length W = width
2. $W = Fs$	W = work F = force s = distance
3. $M = mv$	M = momentum m = mass v = velocity

4. $D = \frac{m}{V}$	D = density m = mass V = volume
5. $v = f\lambda$	v = speed of a wave f = frequency λ = wavelength
6. $y = mx + b$	Slope; y-intercept equation
7. $a^2 + b^2 = c^2$	Pythagorean relation a + b = length of right triangle legs c = length of hypotenuse
8. $P = 2(L + W)$	P = perimeter L = length W = width
9. $A = \frac{1}{2}ab$	A = area of a right triangle a = length of one leg b = other leg
10. $F = ma$	F = force m = mass a = acceleration
11. $\bar{v} = \frac{s}{t}$	$\bar{v}$ = average speed s = distance traveled t = time
12. $a = \frac{v - v_o}{t}$	$v - v_o$ = change in velocity v = final velocity v_o = initial velocity a = acceleration t = time
13. $K.E. = \frac{1}{2}mv^2$	K.E. = Kinetic Energy m = mass v = velocity
14. $P.E. = mgh$	P.E. = Potential Energy m = mass g = acceleration of gravity h = vertical distance mass is lifted

15. $v = v_o + at$	v = final velocity v_o = initial velocity a = acceleration t = time
16. $Ax + By + C = 0$	General equation for a straight line on an X–Y coordinate system.
17. $\frac{x}{a} + \frac{y}{b} = 1$	Intercept equation for an X–Y coordinate system.
18. $x^2 + y^2 = r^2$	General equation for a circle in an X–Y coordinate system: center at origin, radius r.
19. $y = ax^2 + bx + c$	General equation for a parabola with axis parallel to the Y axis (X–Y coordinate system).
20. $h = \frac{ab}{c}$	h = height of a right triangle a = one leg b = the other leg c = hypotenuse
21. $c = 2\pi r$	c = circumference of a circle π = 3.14 r = radius
22. $V = \frac{1}{3} \pi r^2 h$	V = volume of a right circular cone π = 3.14 r = radius of base h = height
23. $V = \frac{4}{3} \pi r^3$	V = volume of a sphere π = 3.14 r = radius
24. $a = \sqrt{(c+b)(c-b)}$	a = one leg of a right triangle b = the other leg c = hypotenuse

Name ______________________ Date ______________________

EQUATIONS TEST

These problems will test your ability to find the value of an unknown variable in an equation. You are not expected to complete the test, so don't race the clock to get done. Work at a steady rate, concentrate on neatness, and strive for accuracy; try to get as many *correct* answers as possible in the time you are given. Calculations and answers will be handed in on notebook paper to be graded, so be sure all work is neat and orderly. Number each answer to avoid confusion.

You will receive two points for each correct answer and two points for each set of calculations that is arranged properly to give the right answer. The teacher will explain how to arrange calculations to earn points. You will receive 4 extra points if your work is neat and orderly.

DIRECTIONS:

On the left side of each column are the equations you will be working with. To the right of each equation is the information that is given to you and the variable that you are to find an answer for. First, get the unknown variable alone on one side of the equals sign, and then solve the equation. Remember, put your calculations and answer to each problem on a separate piece of paper. Round your answers to the nearest hundredth.

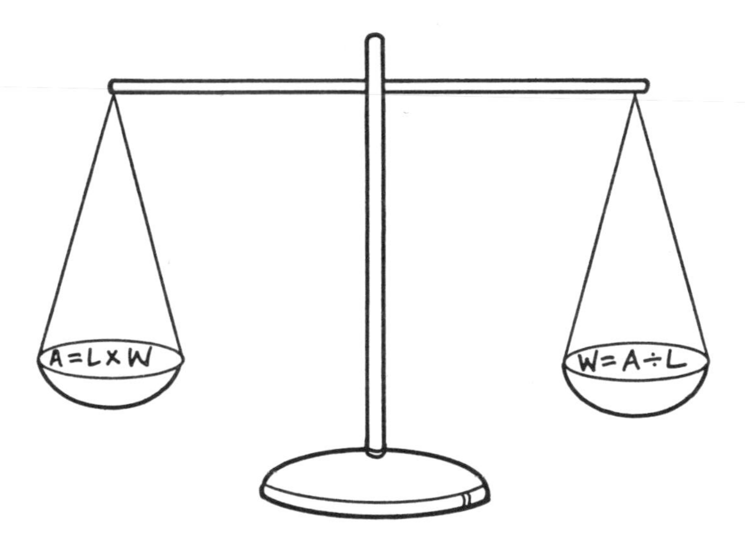

1. $A = L \times W$	Given: $L = 9$, $W = 7$; Find: A	5. $v = f\lambda$	Given: $v = 304.8$, $\lambda = 12.7$; Find: f
2. $W = Fs$	Given: $F = 8.3$, $s = 6.7$; Find: W	6. $y = mx + b$	Given: $y = 521$, $m = 14$, $b = 17$; Find: x
3. $M = mv$	Given: $m = 12.7$, $v = 43.6$; Find: M	7. $a^2 + b^2 = c^2$	Given: $b = 4$, $c = 5$; Find: a
4. $D = \frac{m}{V}$	Given: $m = 74.4$, $V = 12.4$; Find: D	8. $P = 2(L+W)$	Given: $L = 14.2$, $W = 23.6$; Find: P

Name ________________________________ Date ____________________________

EQUATIONS TEST (continued)

9. $A = \frac{1}{2}ab$	Given: $A = 184$, $b = 16$; Find: a	17. $\frac{x}{a} + \frac{y}{b} = 1$	Given: $a = 16$, $b = 8$, $y = 6$; Find: x
10. $F = ma$	Given: $F = 1174.2$, $a = 41.2$; Find: m	18. $x^2 + y^2 = r^2$	Given: $x = 4$, $r = \sqrt{41}$; Find: y
11. $\bar{v} = \frac{s}{t}$	Given: $\bar{v} = 12.0$, $t = 33.5$; Find: s	19. $y = ax^2 + bx + c$	Given: $y = 82$, $a = 6$, $x = 3$, $c = 16$; Find: b
12. $a = \frac{v - v_o}{t}$	Given: $a = 7$, $v_o = 14$, $t = 6$; Find: v	20. $h = \frac{ab}{c}$	Given: $h = 54$, $b = 8$, $c = 4$; Find: a
13. $K.E. = \frac{1}{2}mv^2$	Given: $m = 16$, $v = 7$; Find: K.E.	21. $c = 2\pi r$	Given: $c = 81.64$, $\pi = 3.14$; Find: r
14. $P.E. = mgh$	Given: P.E. $= 540$, $m = 4$, $g = 9$; Find: h	22. $V = \frac{1}{3}\pi r^2 h$	Given: $V = 50$, $r = 5$, $\pi = 3.14$; Find: h
15. $v = v_o + at$	Given: $v = 207$, $v_o = 31$, $a = 8$; Find: t	23. $V = \frac{4}{3}\pi r^3$	Given: $r = 2$, $\pi = 3.14$; Find: V
16. $Ax + By + C = 0$	Given: $A = 4$, $x = 5$, $B = -2$, $y = 16$; Find: C	24. $a = \sqrt{(c+b)(c-b)}$	Given: $b = 3$, $c = 5$; Find: a

Name ______________________________ Date ______________________

EQUATIONS SELF-IMPROVEMENT TEST 1

Equation	Given	Find
1. $A = L \times W$	$L = 8$, $W = 7$	A
2. $W = Fs$	$F = 7.8$, $s = 5.9$	W
3. $M = mv$	$m = 11.8$, $v = 52.2$	M
4. $D = \frac{m}{V}$	$m = 118.4$, $V = 14.8$	D
5. $v = f\lambda$	$v = 508.4$, $\lambda = 16.4$	f
6. $y = mx + b$	$y = 367$, $m = 12$, $b = 19$	x
7. $a^2 + b^2 = c^2$	$b = 6$, $c = 10$	a
8. $P = 2(L + W)$	$L = 15.3$, $W = 25.4$	P
9. $A = \frac{1}{2}ab$	$A = 171$, $b = 18$	a
10. $F = ma$	$F = 984.3$, $a = 38.6$	m
11. $\bar{v} = \frac{s}{t}$	$\bar{v} = 14.0$, $t = 41.7$	s
12. $a = \frac{v - v_o}{t}$	$a = 11$, $v_o = 18$, $t = 4$	v
13. $K.E. = \frac{1}{2}mv^2$	$m = 17$, $v = 8$	$K.E.$
14. $P.E. = mgh$	$P.E. = 792$, $m = 6$, $g = 11$	h
15. $v = v_o + at$	$v = 231$, $v_o = 49$, $a = 7$	t
16. $Ax + By + C = 0$	$A = 4$, $x = 7$, $B = -3$, $y = 12$	C
17. $\frac{x}{a} + \frac{y}{b} = 1$	$a = 6$, $b = 12$, $y = 8$	x
18. $x^2 + y^2 = r^2$	$x = 3$, $r = \sqrt{45}$	y
19. $y = ax^2 + bx + c$	$y = 125$, $a = 5$, $x = 4$, $c = 21$	b
20. $h = \frac{ab}{c}$	$h = 35$, $b = 7$, $c = 5$	a
21. $c = 2\pi r$	$c = 144.44$, $\pi = 3.14$	r
22. $V = \frac{1}{3}\pi r^2 h$	$V = 108$, $r = 6$, $\pi = 3.14$	h
23. $V = \frac{4}{3}\pi r^3$	$r = 3$, $\pi = 3.14$	V
24. $a = \sqrt{(c+b)(c-b)}$	$b = 6$, $c = 10$	a

Name ______________________ Date ______________________

EQUATIONS SELF-IMPROVEMENT TEST 2

Equation	Given	Find
1. $A = L \times W$	$L = 8$, $W = 6$	A
2. $W = Fs$	$F = 7.4$, $s = 9.6$	W
3. $M = mv$	$m = 14.7$, $v = 39.5$	M
4. $D = \frac{m}{V}$	$m = 107.1$, $V = 11.9$	D
5. $v = f\lambda$	$v = 372.6$, $\lambda = 13.8$	f
6. $y = mx + b$	$y = 546$, $m = 16$, $b = 18$	x
7. $a^2 + b^2 = c^2$	$b = 9$, $c = 15$	a
8. $P = 2(L + W)$	$L = 13.8$, $W = 26.3$	P
9. $A = \frac{1}{2}ab$	$A = 180$, $b = 15$	a
10. $F = ma$	$F = 1085.6$, $a = 36.8$	m
11. $\bar{v} = \frac{s}{t}$	$\bar{v} = 16.0$, $t = 38.4$	s
12. $a = \frac{v - v_o}{t}$	$a = 23$, $v_o = 20$, $t = 2$	v
13. $K.E. = \frac{1}{2}mv^2$	$m = 14$, $v = 9$	$K.E.$
14. $P.E. = mgh$	$P.E. = 312$, $m = 3$, $g = 8$	h
15. $v = v_o + at$	$v = 226$, $v_o = 55$, $a = 9$	t
16. $Ax + By + C = 0$	$A = 2$, $x = 8$, $B = -4$, $y = 8$	C
17. $\frac{x}{a} + \frac{y}{b} = 1$	$a = 5$, $b = 10$, $y = 4$	x
18. $x^2 + y^2 = r^2$	$x = 5$, $r = \sqrt{74}$	y
19. $y = ax^2 + bx + c$	$y = 167$, $a = 4$, $x = 5$, $c = 27$	b
20. $h = \frac{ab}{c}$	$h = 62$, $b = 6$, $c = 3$	a
21. $c = 2\pi r$	$c = 106.76$, $\pi = 3.14$	r
22. $V = \frac{1}{3}\pi r^2 h$	$V = 98$, $r = 7$, $\pi = 3.14$	h
23. $V = \frac{4}{3}\pi r^3$	$r = 4$, $\pi = 3.14$	V
24. $a = \sqrt{(c+b)(c-b)}$	$b = 9$, $c = 15$	a

Name ______________________________ Date ______________________

DEGREE-OF-IMPROVEMENT CHART

Name	Self-Improvement Points
	0 5 10 15 20 25 30 35 40 45

Name ______________________________ Date ______________________________

EQUATIONS SELF-IMPROVEMENT TEST 3

Equation	Given	Find
1. $A = L \times W$	L W	A
2. $W = Fs$	F s	W
3. $M = mv$	m = v =	M
4. $D = \frac{m}{V}$	m = V =	D
5. $v = f\lambda$	v = λ =	f
6. $y = mx + b$	y = m = b =	x
7. $a^2 + b^2 = c^2$	b = c =	a
8. $P = 2(L + W)$	L = W =	P
9. $A = \frac{1}{2}ab$	A = b =	a
10. $F = ma$	F = a =	m
11. $\bar{v} = \frac{s}{t}$	$\bar{v}$ = t =	s
12. $a = \frac{v - v_o}{t}$	a = v_o = t =	v
13. $K.E. = \frac{1}{2}mv^2$	m = v =	K.E.
14. $P.E. = mgh$	P.E. = m = g =	h
15. $v = v_o + at$	v = v_o = a =	t
16. $Ax + By + C = 0$	A = x = B = y =	C
17. $\frac{x}{a} + \frac{y}{b} = 1$	a = b = y =	x
18. $x^2 + y^2 = r^2$	x = r =	y
19. $y = ax^2 + bx + c$	y = a = x = c =	b
20. $h = \frac{ab}{c}$	h = b = c =	a
21. $c = 2\pi r$	c = π = 3.14	r
22. $V = \frac{1}{3}\pi r^2 h$	V = r = π = 3.14	h
23. $V = \frac{4}{3}\pi r^3$	r = π = 3.14	V
24. $a = \sqrt{(c+b)(c-b)}$	b = c =	a

EQUATIONS: TEACHER ANSWER SHEET

Equations Test		Self-Improvement Test 1	Self-Improvement Test 2
1.	A = 63	A = 56	A = 48
2.	W = 55.61	W = 46.02	W = 71.04
3.	M = 553.72	M = 615.96	M = 580.65
4.	D = 6	D = 8	D = 9
5.	f = 24	f = 31	f = 27
6.	x = 36	x = 29	x = 33
7.	a = 3	a = 8	a = 12
8.	P = 75.60	P = 81.40	P = 80.20
9.	a = 23	a = 19	a = 24
10.	m = 28.50	m = 25.50	m = 29.50
11.	s = 402	s = 583.80	s = 614.40
12.	V = 56	V = 62	V = 66
13.	K.E. = 392	K.E. = 544	K.E. = 567
14.	h = 15	h = 12	h = 13
15.	t = 22	t = 26	t = 19
16.	c = 12	c = 8	c = 16
17.	x = 4	x = 2	x = 3
18.	y = 5	y = 6	y = 7
19.	b = 4	b = 6	b = 8
20.	a = 27	a = 25	a = 31
21.	r = 13	r = 23	r = 17
22.	h = 1.91	h = 2.87	h = 1.91
23.	V = 33.49	V = 113.04	V = 267.95
24.	a = 4	a = 8	a = 12

Note: Answers are rounded to nearest hundredths.

M-3

MATHEMATICS TREASURE HUNT

Teacher Preview

General Explanation:

Mathematics Treasure Hunt is a problem-solving game that emphasizes basic arithmetic skills. Depending upon the level of skill development in students, this project may be used as a review at the beginning of a school year or as a final project at the end of a year (or semester). Mathematics Treasure Hunt was designed to focus on these areas:

- simplifying fractions
- adding, subtracting, multiplying, and dividing fractions
- changing fractions to percents
- using the distributive property with fractions
- changing improper fractions to mixed numbers
- changing mixed numbers to fractions
- equivalent fractions
- reciprocals
- rounding whole numbers
- rounding decimals
- adding, subtracting, multiplying, and dividing decimals
- changing fractions to decimals
- geometry
- squaring numbers and finding square roots
- cubing numbers
- equations and related sentences
- equalities and inequalities
- lowest common multiples
- greatest common factors
- prime factors
- ratios

Length of Project: 9 hours
Level of Independence: Basic

Goals:

1. To test students' basic skills in mathematics.
2. To provide a game that emphasizes problem solving.
3. To provide motivation for solving problems.

During This Project Students Will:

1. Demonstrate proper use of basic arithmetic skills.
2. Collect data that will be used to make calculations.

Skills:

Collecting data
Listening
Neatness
Identifying problems
Accepting responsibility
Concentration
Controlling behavior
Following project outlines
Individualized study habits
Persistence
Personal motivation
Sense of "quality"
Self-confidence
Addition of fractions
Addition of decimals
Subtraction of fractions
Subtraction of decimals
Multiplication of fractions
Multiplication of decimals
Division of fractions
Division of decimals
Observing
Using equations and related sentences
Accuracy
Solving for an unknown
Basic mathematics skills
Multiplication and addition properties

Handouts Provided:

- "Student Assignment Sheet"
- "Problem Sets" (1–10, 11–20, 21–30, 31–40)
- "Student Answer Sheet for Problem Sets 1–40"
- "Clue Sheet"
- "Treasure Map"
- "Student Answer Sheet for Map Problems"

PROJECT CALENDAR:

HOUR 1: ______	HOUR 2: ______	HOUR 3: ______
The game and its rules are introduced. All work is done in class, so materials are turned in at the end of each hour and handed back at the beginning of each hour. HANDOUTS PROVIDED ANSWERS PROVIDED	Materials are returned and the game begun. Students work on problem sets and gradually collect clues for their maps. PREPARATION REQUIRED	Students continue to work on problem sets and collect clues.
HOUR 4: ______ The game continues.	**HOUR 5: ______** The game continues.	**HOUR 6: ______** The game continues.
HOUR 7: ______ The game continues.	**HOUR 8: ______** The game continues.	**HOUR 9: ______** Students finish their Treasure Map Answer Sheets and turn them in to be graded. STUDENTS TURN IN WORK ANSWERS PROVIDED

Lesson Plans and Notes

HOUR 1: Introduce students to the game and its rules. Tell them that problem sets, answer sheets, clue sheets, and treasure maps will be handed out at the beginning of each hour and collected at the end of each hour. No work is to be done at home. Give students the handouts and explain each one. If time remains after the entire project is thoroughly explained, students may begin solving problems from the first set (simplifying fractions) or this set may be done in class as an example. Materials are handed in at the end of the hour.

HOW THE GAME IS PLAYED

1. Students are given worksheets containing a total of 40 sets of mathematics problems. Each set contains five problems that are similar in difficulty. So, the worksheets have 200 mathematics problems that cover many of the basic skill areas required in upper elementary and junior high mathematics programs.
2. Students work on these problems individually at their desks. When a student completes the first set (five problems) he brings it to a designated place to be checked. If all five answers are correct, he is shown a card that has clue 1 for the treasure hunt written on it. He copies this clue onto his clue sheet, gives the card back, and returns to his seat to begin working on the second set of worksheet problems.
3. If all five answers are *not* correct the student may be told which ones are wrong, or he may simply be told that they are not all correct. He returns to his seat to correctly answer all five problems before being shown a clue card for that set.
4. When the second set of problems is correctly answered, the second treasure hunt clue is awarded, and this process continues until all forty clues have been recorded on a student's "clue sheet."
5. The clues are actually data that will allow students to calculate how far it is from start to finish on their treasure maps. It is left to the students to determine how each piece of information should be used, but the clues are generally presented in the order that they will be needed to calculate the length of all the line segments of the trail, from the beginning to the end. The lengths of these segments are recorded on the "Teacher Answers for the Treasure Hunt Map."
6. Students complete the project when they successfully calculate how far a treasure hunter must travel to follow the map and find the Cave of Treasure Mountain. The object of the game is to find the correct answer, not to finish first. This is not a competition. There is a treasure for *every person* who makes it through the project, regardless of time.

Notes:

- Either prohibit the use of calculators during this project or provide one for every student. It is obviously an unfair advantage for some students to use calculators if others do not. The game will go much faster if calculators are used.
- Some confusion may be expected as the project begins: students won't be sure where to record answers and clues, or which problems to work on next, or what to do when a set of five problems is completed. Much of the confusion can be eliminated by clearly explaining the project, taking the entire class through a sample set of problems, and requiring neatly organized, properly numbered calculations on notebook paper. *Each answer on the calculation paper should be circled.* Calculations should be numbered by problem set and letter: 3-A, 7-C, 19-E, and so forth. These calculations are invaluable tools for discovering if a student knows how to solve a problem, but checking someone's work is time-consuming if the calculations are crammed somewhere on a piece of scrap paper that is cluttered with numbers.

HOUR 2: Give students the problem worksheet(s) and have them begin working on the problem sets. When a student completes a set (five problems), he or she brings it to a grader; if all five answers are correct, that student is given a clue card. The student records the clue on his or her clue sheet and returns the card to the grader. The clue cards are numbered to correspond with set numbers, so if set 6 is correctly completed, clue 6 is awarded. Students work the entire hour to solve problems and collect clues.

Notes:

- Students should put their answers for problem sets 1-40 on the Student Answer Sheet that is provided, to speed up the checking process: it is very important to be able to check answers quickly. Each student's calculations should be numbered and neatly recorded on notebook paper. If copying is suspected, or if you want to see how someone arrived at an answer, ask to see the work. Otherwise, answers can be checked in a matter of seconds.
- Make a set of clue cards, numbered 1 through 40. A 3-by-5-inch notecard makes a good clue card. The clues are provided for you, but they should be transferred to cards so that students can see only one clue at a time.
- It is helpful to have at least one aide—parent, student teacher, or other person—to help check answers and hand out and collect clue cards. One set of clue cards and a teacher answer sheet should be provided for each adult in the room.
- You may also want to consider having one or two "tutors" in the room during this project (parents, adult volunteers, aides, older students, student teachers) to help students who become stumped on particular sets of problems. If a student consistently has incorrect answers for a certain set of problems, assign him or her to spend some time with a tutor so that basic mathematics skills or procedures can be explained.

HOURS 3–8: Students continue to work on their problem sets and gradually collect clues for their treasure maps.

HOUR 9: Students finish their Treasure Map Answer Sheets and turn them in to be graded. Incorrect answers are marked so students can recalculate them. When an answer sheet is 100% correct, the student who completed it is done with the project. Students who finish early are allowed to create their own treasure maps.

Notes:

- The final grading comes when students turn in their Treasure Map Answer Sheets. These sheets are not graded until they are completed. In other words, students do not have each calculation graded individually, as it is finished; answers are graded collectively when all calculations are done.
- When calculating answers to the Treasure Hunt Map, students must round *each* calculation to the nearest thousandth. In some cases two or more calculations are performed to come up with a final line segment distance on the map. Each of these calculations must be rounded to the nearest thousandth to get the answer that is recorded on the Teacher Answer Sheet. You may want to tell students that an answer will be accepted if it is within two thousandths of a mile of the distance given on the answer sheet. To be strict about rounding, however, require that *every* answer to *every* calculation be rounded to the nearest thousandth. This will yield the answers that are provided.

General Notes About This Project:

- This project is designed to be conducted as a class game with individual participants. However, students can work as partners without making significant changes. Mathematics Treasure Hunt is also an excellent independent study project, and it can be used with a portion of a class if other students need individual attention.
- There are a lot of arithmetic problems involved in this project: 40 sets with 5 problems to a set equals 200 problems to be done to obtain 40 treasure hunt clues. The project may be shortened by reducing the number of problems required for each clue.
- You may want to develop a series of homework assignments to go along with the areas of mathematics covered in this project. (They are listed on page 37 under General Explanation.) If skill deficiencies are discovered in certain students, these assignments can be used to help build those skills.
- It is left to the student to decide whether to solve the Treasure Map gradually, as clues are collected, or solve the problem sets in succession and leave the Treasure Map until all 40 clues have been recorded. Some students may prefer one method and some the other.

MATHEMATICS TREASURE HUNT: CLUES FOR TREASURE MAP

These clues should be rewritten on 3-by-5-inch notecards, and each card should be properly numbered. When students earn clues, they are given the appropriate card; they record the clues on their "Clue Sheets" and return the cards to you. Letters with apostrophes (A', B', C'...) are used because there are more than 26 points on the map; after point Z comes point A' (read "A prime").

Note:

$\overline{AB}$ means "line segment AB." This refers to the treasure map and it indicates the distance from point A to point B.

1. $\overline{AB}$ = 1.67 miles.
2. $\overline{AC}$ is 2.09 times longer than $\overline{AB}$.
3. $\overline{CE}$ is 38.72% of $\overline{DF}$.
4. $\overline{EF}$ = 2.71 miles.
5. $\overline{DE}$ = 2.91 miles.
6. The distance from "E" to the *west* end of the bridge is 2.36 miles.
7. The distance from "G" to the *east* end of the bridge is 5.81 miles.
8. The bridge is 0.43 miles long.
9. $\overline{GJ}$ is 1.50 times longer than $\overline{GH}$ + $\overline{JK}$.
10. $\overline{GH}$ = 2.49 miles.
11. $\overline{JK}$ = 2.07 miles.
12. $\overline{JL}$ is $\frac{3}{8}$ of $\overline{JM}$.
12. $\overline{JM}$ = 6.31 miles.
14. $\overline{NO}$ is 1.119 times longer than $\overline{OP}$.
15. $\overline{OP}$ = 2.862 miles.
16. $\overline{LN}$ = $\overline{NO}$.
17. $\overline{QR}$ = 11.041 miles.
18. $\overline{OQ}$ = 3.217 miles.
19. $\overline{RU}$ = 6.213 miles.
20. $\overline{SU}$ = 6.018 miles.
21. $\overline{TU}$ = 2.289 miles.
22. $\overline{VX}$ = 3.29 times $\overline{WX}$.
23. $\overline{TW}$ = 0.342 times $\overline{VX}$.
24. $\overline{WX}$ = 4.08 miles.
25. $\overline{WY}$ = $\frac{1}{4}$ times $\overline{SU}$.

26. $\overline{ZA'}$ = $2\frac{1}{5}$ times $\overline{WY}$.
27. $\overline{WZ}$ = $\overline{ZA'}$.
28. $\overline{ZC'}$ is 27.4% longer than $\overline{ZB'}$.
29. $\overline{ZB'}$ = 5.017 miles.
30. $\overline{D'E'}$ = 10.62 times the length of the rope bridge.
31. $\overline{C'D'}$ = 0.668 times $\overline{D'E'}$.
32. The rope bridge is divided into 24 equal sections.
33. Each section of the rope bridge is divided into 24 mini-sections.
34. Each mini-section of the rope bridge is 0.0008 miles long.
35. $\overline{D'F'}$ = 3.789 miles.
36. $\overline{D'G'}$ = 1.061 times $\overline{D'F'}$.
37. $\overline{G'H'}$ is 92.63% the length of $\overline{D'F'}$ + $\overline{D'G'}$.
38. $\overline{H'I'}$ = $\frac{1}{2}$ $\overline{D'G'}$.
39. I' to the cave entrance = $\frac{1}{13}$ times $\overline{D'I'}$.
40. The distance from the cave entrance to the treasure inside the mountain is 0.012 miles.

Name ______________________________ Date ______________________

MATHEMATICS TREASURE HUNT
Student Assignment Sheet

The object of this project is to find out how far it is from the starting point on the treasure map to the treasure inside the Cave of Treasure Mountain. The only way this distance can be calculated is to gather clues from the teacher: 40 clues are needed to solve the problem. The teacher will not just *give* clues away; each clue will be earned. Here is how the treasure hunt project works:

1. You will receive 4 worksheets of basic mathematics problems. The problems are divided into 40 sets; there are 5 problems in each set.
2. Calculate the answers to all 5 problems in "set 1." Your calculations should be numbered and neatly recorded on notebook paper, with *answers circled.*
3. Record the answers to these 5 problems in the proper spaces on your "Student Answer Sheet."
4. Bring the answer sheet to a classroom grader and have the answers for "set 1" checked.
5. If all 5 answers are correct, you will be given clue card 1. Copy this clue onto your "Treasure Hunt Clue Sheet" and *return the clue card to the grader.* Returning the clue card is important because you do not want everyone to get a free look at it when you had to earn the right to see it.
6. After clue 1 is recorded on the "Treasure Hunt Clue Sheet" you can begin working on the 5 problems of "set 2." When these problems are completed, go through the same process of having your answers checked and then record clue 2.
7. This process continues until you have completed all 40 sets of problems and collected all 40 clues. Use the 40 clues to calculate the distance to the treasure of Treasure Mountain. Use the "Student Answer Sheet for Map Problems" to record map answers.
8. If the grader says that the 5 answers for a set of problems are not all correct, return to your desk to hunt for mistakes. You will not receive a clue for a set of answers until all 5 answers are correct. It should be obvious that it is worth the time to check and recheck answers before going to a grader.
9. If you have difficulty with a certain set, you may be able to spend some time with a tutor who will explain how to work the problems.
10. This project will take several hours to complete. At the end of each hour, hand *all* of the materials in; *it is not fair to copy problems to work on at home!* Also, the goal of the project is to get to Treasure Mountain by calculating how long each segment of the path is. The object of the game is to find the correct answer, not to finish first. This is not a competition: accuracy, neatness, and effort combine to make you successful in the Mathematics Treasure Hunt.

Name ______________________________ Date ____________________

MATHEMATICS TREASURE HUNT: Problem Sets 1–10

1. Simplify each fraction:
   - a) $\frac{4}{16}$
   - b) $\frac{8}{64}$
   - c) $\frac{12}{15}$
   - d) $\frac{12}{32}$
   - e) $\frac{7}{63}$

2. Round each number to the nearest hundred:
   - a) 983
   - b) 2,315
   - c) 76,152
   - d) 98,891
   - e) 206,501

3. Round each number to the nearest tenth:
   - a) 6.49
   - b) 19.61
   - c) 34.653
   - d) 56.119
   - e) 228.6387

4. Add these numbers; simplify your answers:
   - a) $2\frac{1}{2} + 3\frac{1}{3}$
   - b) $7\frac{1}{8} + 3\frac{1}{6}$
   - c) $6\frac{2}{9} + 2\frac{5}{6}$
   - d) $3\frac{2}{5} + 1\frac{1}{8}$
   - e) $12\frac{3}{7} + 16\frac{11}{12}$

5. Add these numbers:
   - a) 13.5 + 6.3
   - b) 12.6 + 14.51
   - c) 26.38 + 19.67
   - d) 68.68 + 41.32
   - e) 109.78 + 98.45

6. Add these numbers:
   - a) 504.2 + 7.62
   - b) 756.6 + 439.734
   - c) 572.5 + 37.04
   - d) 26.613 + 82.874
   - e) .7516 + .0403

7. Subtract these numbers; simplify:
   - a) $5\frac{1}{4} - 3\frac{2}{5}$
   - b) $1\frac{3}{7} - \frac{2}{3}$
   - c) $\frac{10}{13} - \frac{5}{26}$
   - d) $\frac{4}{7} - \frac{12}{21}$
   - e) $3\frac{2}{5} - 1\frac{1}{8}$

8. Subtract these numbers:
   - a) 589.6 − 72.4
   - b) 38.64 − 16.132
   - c) 131.31 − 7.2
   - d) 756.1 − 3.01
   - e) 8.196 − 7.197

9. Subtract these numbers:
   - a) 58.3 − 7.16
   - b) 96.3 − 5.304
   - c) 37.61 − 19.7
   - d) 972.364 − 354.965
   - e) 4391.04 − 2235.443

10. Multiply these numbers; simplify:
    - a) $\frac{3}{4} \times \frac{5}{8}$
    - b) $\frac{1}{4} \times \frac{2}{3}$
    - c) $\frac{7}{8} \times \frac{8}{9}$
    - d) $\frac{3}{4} \times \frac{8}{13}$
    - e) $4 \times \frac{4}{11}$

Name ______________________ Date ______________________

MATHEMATICS TREASURE HUNT: Problem Sets 11–20

11. Multiply these numbers; simplify:
    - a) $5\frac{1}{6} \times \frac{7}{8}$
    - b) $2\frac{1}{4} \times 3\frac{1}{7}$
    - c) $\frac{5}{8} \times \frac{8}{13}$
    - d) $2\frac{9}{10} \times 1\frac{5}{6}$
    - e) $12\frac{2}{13} \times 8\frac{7}{12}$

12. Multiply these numbers:
    - a) 3.14×9.6
    - b) 57.2×3.6
    - c) 672×1.7
    - d) 423.5×9.4
    - e) 55.73×42.96

13. Multiply these numbers; round your answers to the nearest thousandth:
    - a) 34.6×9.12
    - b) $.9301 \times 3.7$
    - c) 5.497×34.5
    - d) $.706 \times .32$
    - e) 3.768×2.659

14. Divide these numbers; simplify:
    - a) $\frac{7}{11} \div \frac{1}{5}$
    - b) $\frac{8}{9} \div \frac{2}{3}$
    - c) $\frac{3}{4} \div \frac{3}{8}$
    - d) $\frac{7}{8} \div \frac{2}{3}$
    - e) $\frac{3}{7} \div \frac{2}{13}$

15. Divide these numbers; simplify:
    - a) $3\frac{2}{5} \div \frac{5}{7}$
    - b) $1\frac{2}{9} \div \frac{3}{4}$
    - c) $4\frac{1}{2} \div 2\frac{1}{4}$
    - d) $4\frac{1}{2} \div 5\frac{3}{7}$
    - e) $9\frac{1}{3} \div 7\frac{1}{6}$

16. Divide these numbers; round to the nearest hundredth:
    - a) $4.16\overline{)48.32}$
    - b) $9.81\overline{)793.4}$
    - c) $37\overline{)5.43}$
    - d) $10.13\overline{)9.4}$
    - e) $17.1\overline{)53.62}$

17. Divide these numbers; round to the nearest thousandth:
    - a) $13.1 \div 1.41$
    - b) $339.1 \div 7.6$
    - c) $43.96 \div 7.7$
    - d) $147 \div 20.61$
    - e) $405.4 \div 17.21$

18. For a circle: $c = 2\pi r$ $\pi = 3.14$ $A = \pi r^2$
    Solve these problems; round to the nearest hundredth:
    - a) $c = 64$; find r
    - b) $r = 12.6$; find c
    - c) $r = 9.4$; find A
    - d) $c = 21.6$; find r
    - e) $A = 28.26$; find r

19. Find the square of each number below:
    - a) 14
    - b) 2.6
    - c) 3.61
    - d) 121
    - e) 12.32

20. Change these fractions to decimals:
    - a) $\frac{3}{25}$
    - b) $\frac{3}{5}$
    - c) $\frac{7}{8}$
    - d) $\frac{3}{4}$
    - e) $\frac{43}{50}$

Name ______________________ Date ______________________

MATHEMATICS TREASURE HUNT: Problem Sets 21–30

21. Change these fractions to percents; round to the nearest tenth:
    a) $\frac{5}{9}$ d) $\frac{14}{15}$
    b) $\frac{22}{40}$ e) $1\frac{2}{3}$
    c) $\frac{37}{50}$

22. For a rectangle: $A = L \times W$
    $P = 2\,(L + W)$
    Solve these problems; round to the nearest hundredth:
    a) $L=32.76$; $W=16.18$; find A
    b) $A=101.32$; $W=2.61$; find L
    c) $L=67.68$; $W=26.17$; find P
    d) $P=58.6$; $L=15.4$; find W
    e) $P=72.9$; $W=4.7$; find L

23. Use the distributive property to rewrite these equations; you do not need to solve the problems:
    a) $(5+\frac{2}{3}) \times \frac{1}{2} = (\_\times\_) + (\_\times\_)$
    b) $(\_+\_) \times \_ = (3\times\frac{1}{4}) + (\frac{1}{8}\times\frac{1}{4})$
    c) $(2+\frac{3}{5}) \times \_ = (\_\times\frac{1}{8}) + (\_\times\frac{1}{8})$
    d) $(5+\frac{3}{10}) \times 1\frac{1}{6} =$
    e) $(9\times\frac{1}{3}) + (\frac{1}{10}\times\frac{1}{3}) =$

24. Solve these square root problems:
    $\pi = 3.14$
    a) $\sqrt{121}$
    b) $\sqrt{169}$
    c) $\sqrt{289}$
    d) $A=\pi r^2$; $A=28.26$; find r
    e) $a^2+b^2=c^2$; $a=9$, $b=12$; find c

25. For a right triangle: $A = \frac{1}{2}\,(h \times b)$
    $c^2 = a^2 + b^2$
    a) $h=5$; $b=6\frac{1}{2}$; find A
    b) $A=26\frac{1}{2}$; $h=4$; find b
    c) $c=20$; $a=12$; find b
    d) $a=15$; $b=20$; find c
    e) $h=16.3$; $b=11.8$; find A

26. Change these improper fractions to mixed numbers:
    a) $\frac{36}{5}$ d) $\frac{98}{3}$
    b) $\frac{64}{7}$ e) $\frac{59}{13}$
    c) $\frac{61}{6}$

27. Use the equation $y = mx + b$ to solve these problems:
    a) $m=13.2$; $x=6.6$; $b=7.3$; find y
    b) $y=38.4$; $m=4.9$; $x=3.2$; find b
    c) $y=62.7$; $m=12.5$; $b=8.1$; find x
    d) $y=53.9$; $x=15.0$; $b=12.2$; find m
    e) $m=2.62$; $x=4.15$; $b=26.031$; find y

28. Use the equation $y = ax^2 + bx + c$ to solve these problems:
    a) $a=11$; $x=5$; $b=9$; $c=56$; find y
    b) $y=156$; $a=3$; $x=4$; $b=7$; find c
    c) $y=273$; $a=12$; $x=3$; $c=13$; find b
    d) $y=311$; $x=5$; $b=6$; $c=20$; find a
    e) $y=426$; $a=4$; $x=6$; $c=48$; find b

29. Change these mixed numbers to improper fractions:
    a) $5\frac{1}{3}$
    b) $11\frac{3}{4}$
    c) $3\frac{2}{9}$
    d) $4\frac{3}{13}$
    e) 8

30. Place the proper sign (=,>,<) between each pair of numbers:
    a) $\frac{2}{7}$ $\frac{10}{35}$
    b) $\frac{5}{8}$ $\frac{15}{32}$
    c) $\frac{3}{11}$ $\frac{13}{44}$
    d) 303 17^2+11
    e) $\frac{4}{9}$ $\frac{45}{109}$

Name ______________________ Date ______________________

MATHEMATICS TREASURE HUNT: Problem Sets 31–40

31. Write two related sentences for each equation below:
    - a) $V = A \times H$
    - b) $d = c \div \pi$
    - c) $A = \pi \times r^2$
    - d) $r = d \div 2$
    - e) $D = T \times S$

32. Write a related sentence for each equation and solve for the unknown:
    - a) $824.18 = 336.4 \times H$
    - b) $24{,}939 = A \times 765$
    - c) $6.5 = 20.41 \div P$
    - d) $16.711 = T \times .983$
    - e) $.3375 = G \div 4$

33. Write the reciprocal for each number below:
    - a) $\frac{3}{8}$
    - b) $\frac{10}{3}$
    - c) 17
    - d) $\frac{2}{9}$
    - e) $3\frac{1}{2}$

34. Find the cube of each number below:
    - a) 2
    - b) 3
    - c) 8
    - d) 2.1
    - e) 3.2

35. Solve each equation below:
    - a $V = \frac{3}{4} \times \frac{5}{8}$
    - b) $V = 5\frac{1}{6} \times \frac{7}{8}$
    - c) $V = 2\frac{9}{10} \times 1\frac{5}{6}$
    - d) $r = \frac{7}{11} \div \frac{1}{5}$
    - e) $r = 4\frac{1}{2} \div 5\frac{3}{7}$

36. Write equivalent fractions for the fractions below:
    - a) $\frac{5}{9} = \frac{15}{\square}$
    - b) $\frac{10}{13} = \frac{\square}{117}$
    - c) $\frac{12}{5} = \frac{\square}{25}$
    - d) $\frac{21}{1} = \frac{126}{\square}$
    - e) $3\frac{1}{6} = \frac{\square}{24}$

37. Find the lowest common multiple for each pair of numbers below:
    - a) 7; 8
    - b) 10; 4
    - c) 3; 13
    - d) 9; 7
    - e) 3; 15

38. Find the greatest common factor for each pair of numbers below:
    - a) 28; 21
    - b) 51; 34
    - c) 56; 20
    - d) 39; 27
    - e) 49; 91

39. Find the prime factors for each number below:
    - a) 12
    - b) 16
    - c) 18
    - d) 39
    - e) 180

40. Make the units the same, then write each ratio below as a fraction:
    - a) 37 seconds to 1 hour
    - b) 57 seconds to 1 day
    - c) 5 inches to 5 yards
    - d) 17 inches to 1 mile (5280 feet/mile)
    - e) 25 seconds to 2 weeks

Name ______________________ Date ______________________

MATHEMATICS TREASURE HUNT: STUDENT ANSWER SHEET FOR PROBLEM SETS 1-40

Set No.	Problem A	Problem B	Problem C	Problem D	Problem E
1.					
2.					
3.					
4.					
5.					
6.					
7.					
8.					
9.					
10.					
11.					
12.					
13.					
14.					
15.					
16.					
17.					
18.					
19.					
20.					
21.					
22.					
23.					
24.					
25.					
26.					
27.					
28.					
29.					
30.					
31.					
32.					
33.					
34.					
35.					
36.					
37.					
38.					
39.					
40.					

Name ______________________ Date ______________________

MATHEMATICS TREASURE HUNT
Clue Sheet

Record each clue on the proper line below; you will get one clue for every set of problems that you correctly complete. *Be very careful to copy the clues accurately!*

Clue 1	
Clue 2	
Clue 3	
Clue 4	
Clue 5	
Clue 6	
Clue 7	
Clue 8	
Clue 9	
Clue 10	
Clue 11	
Clue 12	
Clue 13	

Name ____________________ Date ____________________

MATHEMATICS TREASURE HUNT
Clue Sheet (continued)

Clue 14	
Clue 15	
Clue 16	
Clue 17	
Clue 18	
Clue 19	
Clue 20	
Clue 21	
Clue 22	
Clue 23	
Clue 24	
Clue 25	
Clue 26	
Clue 27	
Clue 28	
Clue 29	
Clue 30	
Clue 31	
Clue 32	
Clue 33	
Clue 34	
Clue 35	
Clue 36	
Clue 37	
Clue 38	
Clue 39	
Clue 40	

Name ______________________________ Date ______________________

TREASURE MAP

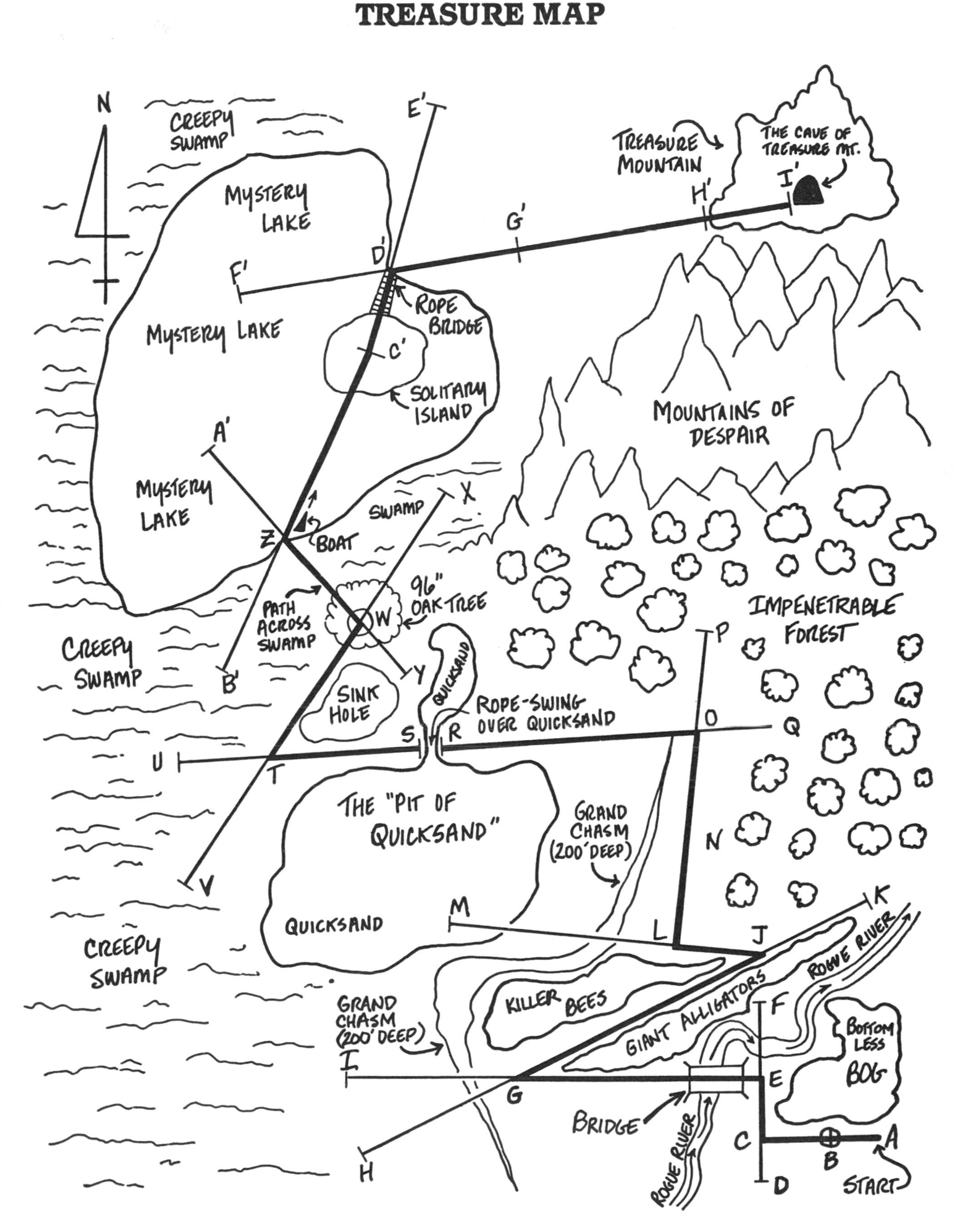

Name ______________________ Date ______________________

MATHEMATICS TREASURE HUNT
Student Answer Sheet for Map Problems

Use the clues that you have recorded on the "Treasure Hunt Clue Sheet" to calculate the distance of each line segment listed below. Put your calculations on notebook paper. The sum of these line segments equals the total distance from "start" to "treasure" on the treasure map. Round all answers to the nearest *thousandth* of a mile, and record them below:

	Line Segment		Distance
1.	$\overline{AC}$	=	______________
2.	$\overline{CE}$	=	______________
3.	$\overline{EG}$	=	______________
4.	$\overline{GJ}$	=	______________
5.	$\overline{JL}$	=	______________
6.	$\overline{LO}$	=	______________
7.	$\overline{OT}$	=	______________
8.	$\overline{TW}$	=	______________
9.	$\overline{WZ}$	=	______________
10.	$\overline{ZC'}$	=	______________
11.	$\overline{C'D'}$	=	______________
12.	$\overline{D'I'}$	=	______________
13.	Distance from I' to the cave entrance	=	______________
14.	Distance from the cave entrance to the treasure	=	______________
	TOTAL	=	______________

MATHEMATICS TREASURE HUNT: TEACHER ANSWERS FOR PROBLEM SETS 1-40

Set No.	Problem A	Problem B	Problem C	Problem D	Problem E
1.	$\frac{1}{4}$	$\frac{1}{8}$	$\frac{4}{5}$	$\frac{3}{8}$	$\frac{1}{9}$
2.	1,000	2,300	76,200	98,900	206,500
3.	6.5	19.6	34.7	56.1	228.6
4.	$5\frac{5}{6}$	$10\frac{7}{24}$	$9\frac{1}{18}$	$4\frac{21}{40}$	$29\frac{29}{84}$
5.	19.8	27.11	46.05	110.00	208.23
6.	511.82	1196.334	609.54	109.487	0.7919
7.	$1\frac{17}{20}$	$\frac{16}{21}$	$\frac{15}{26}$	0	$2\frac{11}{40}$
8.	517.2	22.508	124.11	753.09	0.999
9.	51.14	90.996	17.91	617.399	2155.597
10.	$\frac{15}{32}$	$\frac{1}{6}$	$\frac{7}{9}$	$\frac{6}{13}$	$1\frac{5}{11}$
11.	$4\frac{25}{48}$	$7\frac{1}{14}$	$\frac{5}{13}$	$5\frac{19}{60}$	$104\frac{25}{78}$
12.	30.144	205.92	1142.4	3980.9	2394.1608
13.	315.552	3.441	189.647	0.226	10.019
14.	$3\frac{2}{11}$	$1\frac{1}{3}$	2	$1\frac{5}{16}$	$2\frac{11}{14}$
15.	$4\frac{19}{25}$	$1\frac{17}{27}$	2	$\frac{63}{76}$	$1\frac{13}{43}$
16.	11.62	80.88	0.15	0.93	3.14
17.	9.291	44.618	5.709	7.132	23.556
18.	$r=10.19$	$c=79.13$	$A=277.45$	$r=3.44$	$r=3$
19.	196	6.76	13.0321	14,641	151.7824
20.	0.12	0.6	0.875	0.75	0.86
21.	55.6%	55.0%	74.0%	93.3%	166.7%
22.	$A=530.06$	$L=38.82$	$P=187.7$	$W=13.9$	$L=31.75$
23.	$(5+\frac{2}{3})\times\frac{1}{2}=$ $(5\times\frac{1}{2})+(\frac{2}{3}\times\frac{1}{2})$	$(3+\frac{1}{8})\times\frac{1}{4}=$ $(3\times\frac{1}{4})+(\frac{1}{8}\times\frac{1}{4})$	$(2+\frac{3}{5})\times\frac{1}{8}=$ $(2\times\frac{1}{8})+(\frac{3}{5}\times\frac{1}{8})$	$(5+\frac{3}{10})\times 1\frac{1}{6}=$ $(5\times 1\frac{1}{6})+(\frac{3}{10}\times 1\frac{1}{6})$	$(9\times\frac{1}{3})+(\frac{1}{10}\times\frac{1}{3})=$ $(9+\frac{1}{10})\times\frac{1}{3}$
24.	11	13	17	3	15
25.	$A=16\frac{1}{4}$	$b=13\frac{1}{4}$	$b=16$	$c=25$	$A=96.17$
26.	$7\frac{1}{5}$	$9\frac{1}{7}$	$10\frac{1}{6}$	$32\frac{2}{3}$	$4\frac{7}{13}$
27.	$y=94.42$	$b=22.72$	$x=4.368$	$m=2.78$	$y=36.904$
28.	$y=376$	$c=80$	$b=50.667$	$a=10.44$	$b=39$
29.	$\frac{16}{3}$	$\frac{47}{4}$	$\frac{29}{9}$	$\frac{55}{13}$	$\frac{8}{1}$
30.	$=$	$>$	$<$	$>$	$>$
31.	$A=V\div H$; $H=V\div A$	$c=D\times\pi$; $\pi=c\div D$	$r^2=A\div\pi$; $\pi=A\div r^2$	$D=r\times 2$; $2=D\div r$	$T=D\div S$; $S=D\div T$
32.	$H=824.18\div 336.4$; $H=2.45$	$A=24{,}939\div 765$; $A=32.6$	$P=20.41\div 6.5$; $P=3.14$	$T=16.711\div .986$; $T=17$	$G=.3375\times 4$; $G=1.35$
33.	$\frac{8}{3}$	$\frac{3}{10}$	$\frac{1}{17}$	$\frac{9}{2}$	$\frac{2}{7}$
34.	8	27	512	9.261	32.768
35.	$V=\frac{15}{32}$	$V=4\frac{25}{48}$	$V=5\frac{19}{60}$	$r=3\frac{2}{11}$	$r=\frac{63}{76}$
36.	$\frac{15}{27}$	$\frac{90}{117}$	$\frac{60}{25}$	$\frac{126}{6}$	$\frac{76}{24}$
37.	56	20	39	63	15
38.	7	17	4	3	7
39.	$2\times 2\times 3$	$2\times 2\times 2\times 2$	$2\times 3\times 3$	3×13	$2\times 2\times 3\times 3\times 5$
40.	$\frac{37}{3600}$	$\frac{57}{86{,}400}$	$\frac{5}{180}=\frac{1}{36}$	$\frac{17}{63{,}360}$	$\frac{25}{1{,}209{,}600}=\frac{1}{48{,}384}$

TEACHER ANSWERS FOR THE TREASURE HUNT MAP

(These answers are to be calculated from the clues students earn by solving problem sets 1–40.)

Line segment		Distance
$\overline{AC}$	=	3.490 miles
$\overline{CE}$	=	2.176 miles
$\overline{EG}$	=	7.740 miles
$\overline{GJ}$	=	6.840 miles
$\overline{JL}$	=	2.366 miles
$\overline{LO}$	=	6.406 miles
$\overline{OT}$	=	11.748 miles
$\overline{TW}$	=	4.591 miles
$\overline{WZ}$	=	3.311 miles
$\overline{ZC'}$	=	6.392 miles
$\overline{C'D'}$	=	3.271 miles
$\overline{D'I'}$	=	13.263 miles
I' to the cave entrance	=	1.021 miles
Cave entrance to the treasure	=	0.012 miles
TOTAL	=	72.627 miles

M-4

SECTIONS AND ACRES

Teacher Preview

Length of Project: 8 hours
Level of Independence: Basic
Goals:

1. To introduce students to land surveying.
2. To combine mathematics with geography.
3. To place emphasis on problem-solving skills.
4. To show students that mathematics can be applied to real world problems.

During This Project Students Will:

1. Study the process of land division (sections and acres).
2. Solve land division problems.

Skills:

Listening
Neatness
Concentration
Working with fractions
Working with decimals
Basic mathematics skills
Simple equations
Related sentences
Accuracy
Solving for an unknown
Observing
Sense of "quality"

Handouts Provided:

- "Dividing Land into Blocks, Townships, and Sections"
- "Sections Worksheet"
- "Acres Worksheet"
- "Problem Solving–Sections and Acres"
- "Sections and Acres Test"

PROJECT CALENDAR:

HOUR 1: ______	HOUR 2: ______	HOUR 3: ______
The most common method of dividing land in America—into blocks, townships, sections, and acres—is introduced. PREPARATION REQUIRED HANDOUTS PROVIDED	Students complete a worksheet on *sections* in class. HANDOUT PROVIDED STUDENTS TURN IN WORK ANSWER SHEET PROVIDED	Students complete a worksheet on *acres* in class. RETURN STUDENT WORK HANDOUT PROVIDED STUDENTS TURN IN WORK ANSWER SHEET PROVIDED
HOUR 4: ______	**HOUR 5: ______**	**HOUR 6: ______**
A problem-solving handout is distributed and students work on it in class. RETURN STUDENT WORK HANDOUT PROVIDED STUDENTS TURN IN WORK ANSWER SHEET PROVIDED	Students are given more problems (teacher-made) to solve; these are worked on in class. RETURN STUDENT WORK PREPARATION REQUIRED STUDENTS TURN IN WORK	Procedures for solving sections and acres problems are reviewed. RETURN STUDENT WORK PREPARATION REQUIRED
HOUR 7: ______	**HOUR 8: ______**	**HOUR 9: ______**
Students take a test on sections and acres. HANDOUT PROVIDED ANSWER SHEET PROVIDED	Tests are returned, followed by a discussion of the importance of accurate land division and what was learned during this project. RETURN STUDENT WORK	

Lesson Plans and Notes

HOUR 1: Introduce students to the most common way of dividing and subdividing land in America. Land division illustrations have been provided following these lesson plans; they are designed as handouts but can be used just for your reference. Beginning with the equator and prime meridian, the earth is systematically broken down into smaller and smaller units. Surveyors usually work with areas of one square mile and less, which are referred to as sections and fractions of sections. These areas are commonly measured in acres. Small fractions of sections are often measured in square feet. Detailed notes about land division are provided here for your information:

1. The earth has two points from which all measurement begins: the north pole and the south pole.
2. Lines that run between the north pole and the south pole are called lines of longitude, or meridians.
3. The meridian that runs through Greenwich, England, is called the prime meridian; east and west measurement begins from it.
4. Lines that run perpendicular to the prime meridian are called lines of latitude.
5. The latitude line that divides the globe into two equal parts is called the equator; north and south measurement begins from it.
6. The world is subdivided from the prime meridian and equator by lines called principal meridians (longitude) and base lines (latitude). There are thirty-one pairs, or sets, of these principal meridians and base lines in the continental United States and three pairs in Alaska, from which all U.S. surveys begin. They were established by the U.S. Geologic Survey on geographically convenient lines that avoided mountains, forests, swamps, and lakes. These lines were not necessarily established on precise lines of longitude or latitude. Rather, the lines were created where it is easiest to see and measure long distances. Consequently, there is no standard distance between them.
7. Between each set of principal meridians and base lines, land is divided into squares that are twenty-four miles on each side. The lines that form these squares are called guide meridians (longitude) and standard parallels (latitude).
8. Each twenty-four mile square is divided into squares that are six miles on a side; these squares are called townships and the twenty-four mile square is called a block of townships. One block contains sixteen townships.
9. An east-west row of townships in a block is called a tier. A north-south column of townships is called a range.

10. The north-south lines that identify townships are called range lines (longitude).
11. The east-west lines that identify townships are called townships lines (latitude).
12. Each township is divided into squares that are one mile on a side; these squares are called sections.
13. A township is located by establishing its southwest corner and measuring exactly six miles east and six miles north. The sections are then "boxed in" by measurement from these lines.
14. Sections may be divided into half, quarter, sixteenth, sixty-fourth, or even smaller subdivisions.

    Note:

    - In the real world these "squares" cannot have four equal sides because, in the northern hemisphere, longitude lines are all converging to the same point (the north pole). Therefore the north side of each square must be shorter than the south side. This is compensated for in surveying, but should be ignored in this activity. You may explain in class that it happens, but don't try to compensate mathematically. Assume everything is square.

HOUR 2: Give students the "Sections Worksheet." Have students work the problems on this sheet in class after a discussion of the concepts involved. The worksheets are handed in and graded.

HOUR 3: Return graded worksheets from Hour 2. Give students the "Acres Worksheet" and have them work the problems on this sheet in class after a discussion of the concepts involved. Place emphasis on using related sentences with the equation $A = W \times L$. The worksheets are handed in and graded.

HOUR 4: Return graded worksheets from Hour 3. Give students the "Problem Solving—Sections and Acres" handout that contains seven problems. These problems are also done in class so that students who have difficulty can be helped immediately. Students turn in their work at the end of the hour.

HOUR 5: Return graded worksheets from Hour 4. Give students more problems to solve. These problems are *not* provided because it is important that their difficulty correspond with the capabilities of the class. This gives you the opportunity to create difficult, challenging problems or simple problems that reinforce basic skills. Also, don't neglect the possibility of student-created problems. Students turn in their work at the end of the hour.

HOUR 6: Return graded problems from Hour 5. Use the rest of the hour to review for the test.

HOUR 7: Test.

HOUR 8: Return the tests and discuss them. Then turn the discussion to the importance of accurate land division in America. Why would land owners, bankers, realtors, and builders be concerned with precise location of property corners and calculation of land areas? Is an understanding of this process valuable to anyone other than these people? Introduce students to professions that deal with the division of land: surveyors, civil engineers, land developers, realtors, and farmers, among others. End the discussion with a brief look at the kinds of mathematics projects that the study of sections and acres can lead to: point location, geometry, land surveying, mapping, and investment (money or land-cost) problems.

General Note About This Project:

- There are a number of ways to expand this project. For example, supply maps, plats, or other scale drawings and ask students to "survey" them with protractors and rulers; or, ask them to calculate how many "acres" are in specified areas of each map. Bring a surveyor into the classroom for a talk, or take students outside to see a real survey being conducted. You can also coordinate this project with other projects or courses, such as geography, history, or geometry.

Name ______________________________ Date ______________________________

DIVIDING LAND INTO BLOCKS, TOWNSHIPS, AND SECTIONS

In the early days of this country, explorers, pioneers, and settlers found a land without man-made boundaries. Territories were defined by natural boundaries, such as rivers, mountains, forests, lakes, and other geographic features. Gradually, though, it became necessary to divide the land into areas of uniform size, so it could be bought, sold, fenced, built upon, and occupied. Soon, a person could not sell a piece of land without supplying a document that told precisely where the land was located and how big it was. Such a land description was based upon points and lines that were established by the United States government. This handout is designed to explain and show the system that was developed, over two hundred years ago, to divide and subdivide the land of America.

The first step was to establish a set of lines at various convenient places across the continent (and in Alaska) from which to measure all the rest of the land. These major lines, which are used to begin all land surveys, were called *principal meridians* and *base lines.* Principal meridians run north and south and are lines of longitude. Base lines run east and west and are lines of latitude. Between these lines, over most of the United States, the land was divided into squares twenty-four miles on a side. Each of these squares was called a *block.* The sides of a block that ran north and south were called *guide meridians* and the sides that ran east and west were called *standard parallels.*

Each block was further subdivided into sixteen *townships.* A township was a square piece of land, six miles on a side. The sides of a township that ran north and south were called *range lines* and those that ran east and west were called *township lines.*

Of course, the townships were then subdivided into yet smaller pieces of land, this time into thirty-six squares that were one mile on a side. These squares were called *sections;* a section is one square mile of land. Quite often, during pioneer days, land was sold by the section, although it became common to divide a section into four equal parts before selling the land. These subdivisions were called *quarter sections* because each was one quarter of an entire section.

Study the drawings on this handout to better acquaint yourself with the way land is divided. Notice particularly that the measurement for locating blocks, townships, and sections always begins in the southwest corner. This is done so that all surveys are uniform.

Name ______________________________ Date ______________________________

DIVIDING LAND INTO BLOCKS, TOWNSHIPS, AND SECTIONS (continued)

Locating Blocks of Land

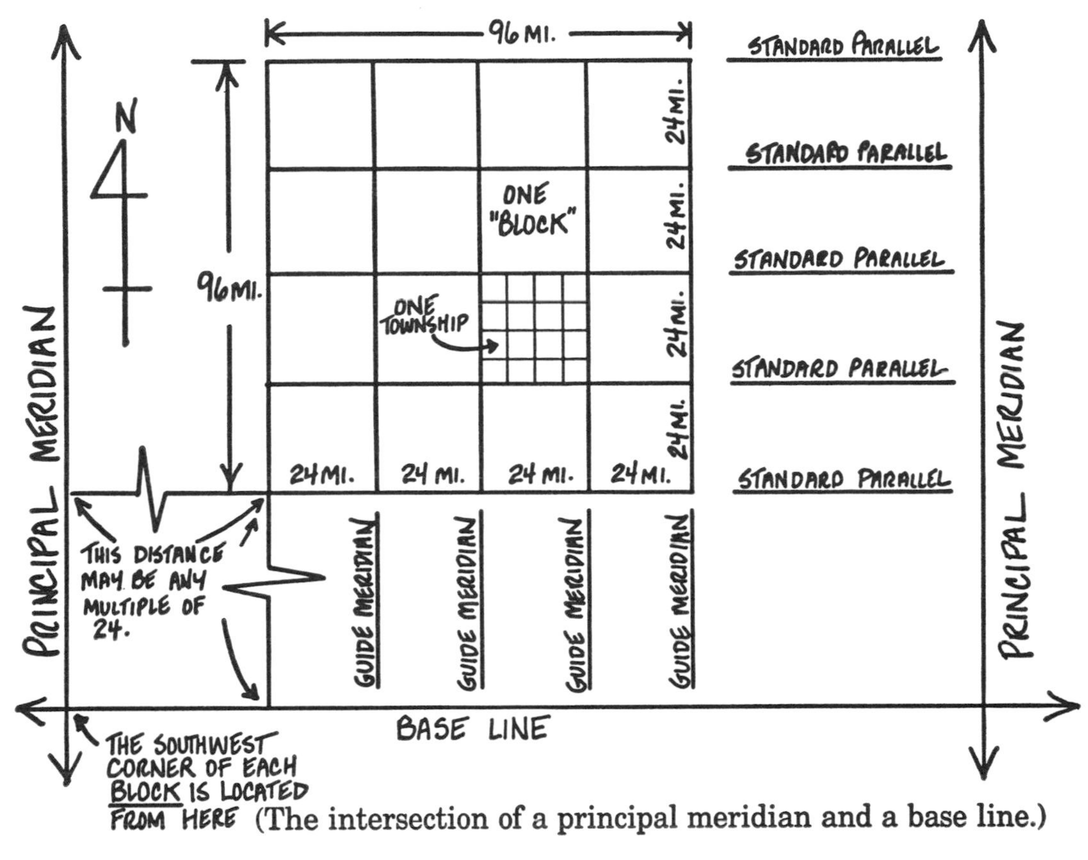

(The intersection of a principal meridian and a base line.)

One Block

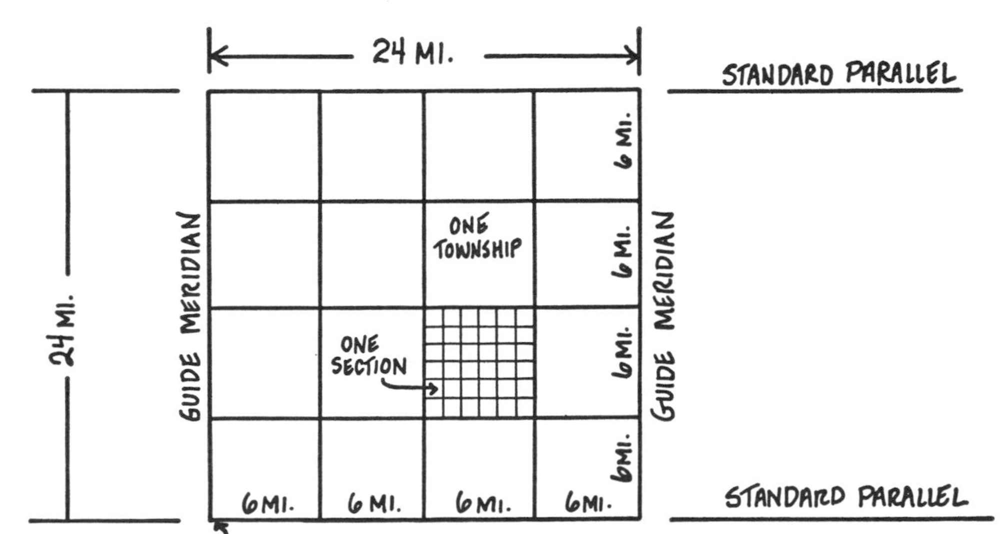

(The intersection of a guide meridian and a standard parallel.)

Name ______________________ Date ______________________

DIVIDING LAND INTO BLOCKS, TOWNSHIPS, AND SECTIONS (continued)

One Township

TOWNSHIP LINE (EXAMPLE: TWP. 9 NORTH)

RANGE LINE (EXAMPLE: RANGE 8 WEST)

6	5	4	3	2	1	1 MI.
7	8	9	10	11	12	1 MI.
18	17	16	15	14	13	1 MI.
19	20	21	22	ONE SECTION 23	24	1 MI.
30	29	28	27	1/4 SECTION 26	25	1 MI.
31 1 MI.	32 1 MI.	33 1 MI.	34 1 MI.	35 1 MI.	36 1 MI.	1 MI.

RANGE LINE (EXAMPLE: RANGE 7 WEST)

TOWNSHIP LINE (EXAMPLE: TWP. 8 NORTH)

THE SOUTHWEST CORNER OF EACH SECTION IS LOCATED FROM HERE (The intersection of a range line and a township line.)

Name ______________________ Date ______________________

SECTIONS WORKSHEET

Most land transactions today involve much less than one section (one square mile). Originally, after the four corners of a section were marked, surveyors would divide it into four equal parts, called quarter sections. Later, each quarter section would be further subdivided as the land was cleared and more people moved into a territory. These subdivisions were called eighth, sixteenth, and sixty-fourth sections, depending on what portion of a section they represented.

This worksheet shows how a section of land is divided into smaller units. Study the drawings, then answer the questions.

1. $\frac{1}{4}$ section = __________ acres.
   $\frac{1}{4}$ section is __________ ft long __________ ft wide

2. $\frac{1}{8}$ section = __________ acres.
   $\frac{1}{8}$ section is __________ ft long __________ ft wide.

3. $\frac{1}{16}$ section = __________ acres.
   $\frac{1}{16}$ section is __________ ft long __________ ft wide.

4. $\frac{1}{64}$ section = __________ acres.
   $\frac{1}{64}$ section is __________ ft long __________ ft wide.

FRACTIONS OF A SECTION

1/16 1/64

1/4 1/8

Name ____________________ Date ____________________

ACRES WORKSHEET

It is common to hear people refer to the size of a piece of land in *acres* rather than as a fraction of a section. This is because many transactions involve very small parcels of land, much smaller than a sixty-fourth or even a one-hundred-twenty-eighth of a section: there are 640 acres in one section. An acre has no specific shape, but it contains a certain number of square feet: 43,560. Knowing this, a piece of land can be rectangular or irregular in shape and its size in acres can still be calculated.

Study each problem below and solve it based upon the information that is given to you.

AREA = LENGTH × WIDTH
LENGTH = AREA ÷ WIDTH
WIDTH = AREA ÷ LENGTH
} RELATED SENTENCES

One acre = 43,560 ft^2 (square feet)

Write the correct related sentence needed to solve the problem. Substitute numbers into the sentence and solve.

1. Find the area of this piece of land. Is it more or less than one acre?

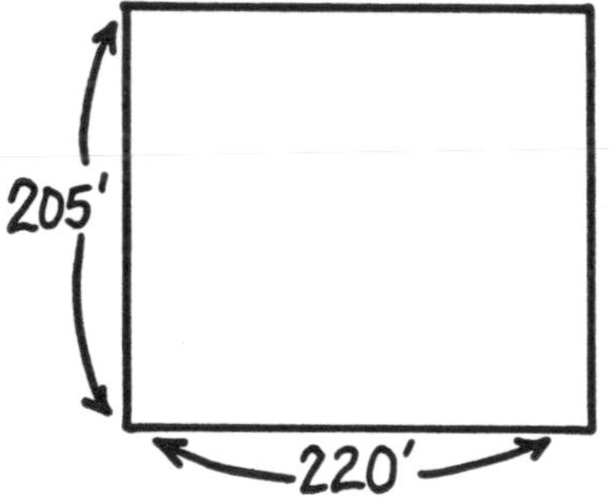

2. What does the width of this piece of land have to be to make one acre?

3. What is the length of this two-acre piece of land?

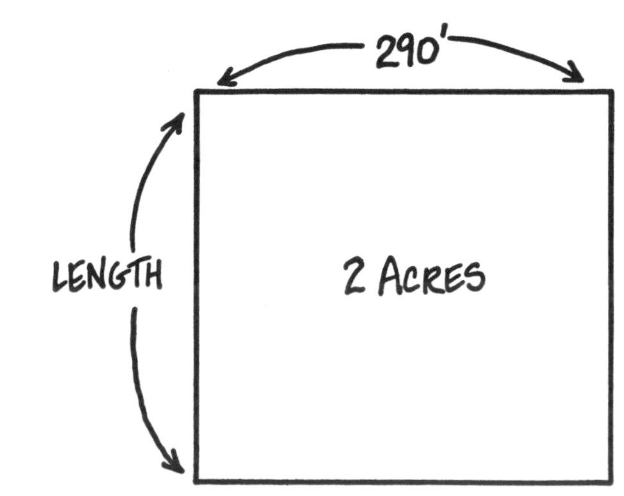

Name ________________________________ Date ____________________________

PROBLEM SOLVING—SECTIONS AND ACRES

Now that you have an understanding of sections and acres, you can solve problems about land division and the cost of buying land. This ability may come in handy when you start thinking about the possibility of buying land for a home, farm, or business enterprise.

Before beginning, write two *related division sentences* for the equation below:

Total Price = (Price per acre) × (number of acres)

1. ____________________ = ____________________ ÷ ____________________

2. ____________________ = ____________________ ÷ ____________________

Directions: For each of the following problems decide which equation you need to use; substitute what you know into the equation and solve for the unknown.

3. Dave wants to buy a $\frac{1}{4}$ section of forest land. The price of the land he wants to buy is $450 per acre. How much will it cost him to buy the land?

4. Dave can't convince the bank to lend him that much money. They will let him borrow $54,000. How many acres can he buy with his loan?

5. Dave owns a piece of land that is $\frac{1}{8}$ mile wide and $\frac{1}{4}$ mile long. He is going to sell it for $675 per acre. How much will he get for his land?

6. With the money he gets for his land and his bank loan, how many acres can Dave buy?

7. How much more money does he need to buy the whole $\frac{1}{4}$ section?

Name ______________________________ Date ______________________

SECTIONS AND ACRES TEST

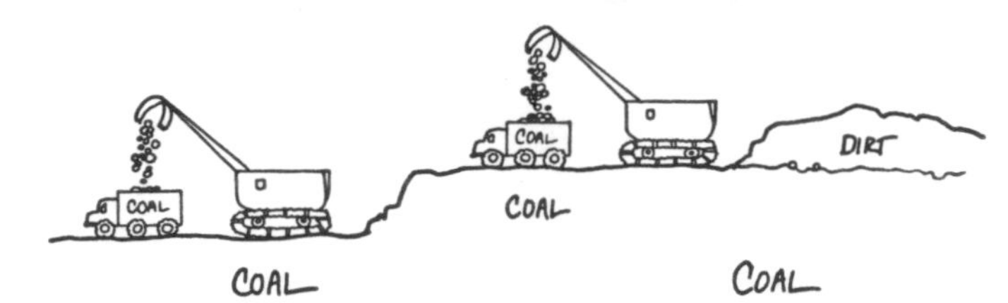

For each problem, list all important information on a separate piece of paper, record the proper equation if one is needed, make a drawing and calculate the answer. Circle your answer and be neat. Answers are *wrong* if they are without units.

1. You own $\frac{1}{8}$ of a section and cut and sell firewood from your land for $40 a cord. From your well-managed forest you can harvest about one cord of wood per acre per year.

   A. How many acres do you own? ______________________

   B. How many cords can you harvest from your land in a year? ______________________

   C. How much money can you take in each year by selling firewood? ______________________

2. You have $15,000 to buy a piece of land. The land you are looking at costs $480,000 for $\frac{1}{2}$ of a section.

   A. How much does this land cost per acre? ______________________

   B. How many acres can you buy? ______________________

   C. What fraction of a section will your purchase be? ______________________

3. You bought three acres of land for $1,400. There are 43,560 ft^2 in an acre.

   A. How much did it cost per acre? ______________________

   B. How many square feet of land do you own? ______________________

   C. Your land is 415 feet long; how wide is it? ______________________

4. A coal company discovered a vein of coal 15 feet below the surface. Before it can be mined, the soil above the coal must be removed (strip mining). The width of the vein is $\frac{1}{16}$ of a mile, the length is $\frac{1}{2}$ mile, and the depth is 100 feet. There are 5,280 feet in a mile.

   A. What is the length of the vein in feet? ______________________

   B. What is the width of the vein in feet? ______________________

   C. How many square feet of land area will be dug up when they mine the coal? ______

   D. How many acres does this equal? ______________________

   E. How many *cubic feet* of soil must be removed to get to the coal? ______________________

   F. How many *cubic feet* of coal are in the vein? ______________________

EXTRA CREDIT: How many *cubic yards* of coal are in the vein? ______________________

TEACHER ANSWER SHEET FOR SECTIONS AND ACRES HANDOUTS

Sections

1. 160 acres
   (2,640 ft long × 2,640 ft wide)
2. 80 acres
   (2,640 ft long × 1,320 ft wide)
3. 40 acres
   (1,320 ft long × 1,320 ft wide)
4. 10 acres
   (660 ft long × 660 ft wide)

Acres

1. 45,100 ft^2
   (more than an acre)
2. 122.7 ft
3. 300.4 ft

Problem Solving–Sections and Acres

1. Price per acre = Total price ÷ Number of acres
2. Number of acres = Total price ÷ Price per acre
3. $72,000
4. 120 acres
5. $13,500
6. 150 acres
7. $4,500

TEACHER ANSWER SHEET FOR SECTIONS AND ACRES TEST

1. A. 80 acres
   B. 80 cords
   C. $3,200
2. A. $1,500
   B. 10 acres
   C. 1/64 section
3. A. $466.67
   B. 130,680 ft^2
   C. 314.9 ft
4. A. 2,640 ft
   B. 330 ft
   C. 871,200 ft^2
   D. 20 acres
   E. 13,068,000 ft^3
   F. 87,120,000 ft^3

EXTRA CREDIT: 3,226,666.6 yd^3

M-5

PROTRACTOR–COMPASS–RULER

Teacher Preview

Length of Project: 13 hours
Level of Independence: Basic
Goals:

1. To prepare students to make quality posters and other visual material for later projects.
2. To teach the skills needed to
   a. Bisect line segments and angles.
   b. Draw parallel lines.
   c. Inscribe and circumscribe triangles with circles.
   d. Locate points.
3. To introduce students to concepts of geometry that can be expanded upon in later courses or used in such specialized projects as land surveying or mapping.

During This Project Students Will:

1. Demonstrate correct use of protractors, compasses, and rulers.
2. Locate points in a plane from a base line, when given angles and distances.
3. Correctly use terms that relate to plane geometry.

Skills:

Drawing straight lines
Drawing parallel lines
Using a compass
Using a protractor
Using a ruler
Using a straightedge
Neatness
Accuracy
Linear measurement
Angular measurement
Concentration
Controlling behavior
Individualized study habits
Persistence
Taking care of materials
Personal motivation
Sense of "quality"
Self-confidence
Listening
Drawing/sketching/graphing
Point location
Observing

Handouts Provided:

- "Protractor–Compass–Ruler: Terms and Definitions"
- "Point Location Worksheet"
- "Protractor–Compass–Ruler: Final Test"

PROJECT CALENDAR:

HOUR 1: ____________	HOUR 2: ____________	HOUR 3: ____________
A vocabulary handout that will be used throughout the project is distributed, followed by an introduction to equipment; students draw and measure line segments, circles, and parallel lines. HANDOUT PROVIDED NEED SPECIAL MATERIALS	Students are taught how to bisect line segments and angles. PREPARATION REQUIRED NEED SPECIAL MATERIALS	Students work on a teacher-made worksheet of line segments and angles to be bisected. PREPARATION REQUIRED NEED SPECIAL MATERIALS
HOUR 4: ____________	**HOUR 5:** ____________	**HOUR 6:** ____________
Students are taught how to circumscribe a triangle with a circle, and they practice on a teacher-made worksheet. PREPARATION REQUIRED NEED SPECIAL MATERIALS	Students are taught how to inscribe a triangle with a circle, and they practice on a teacher-made worksheet. PREPARATION REQUIRED NEED SPECIAL MATERIALS	Students are taught how to measure angles with a protractor. A list of vocabulary words from the Terms and Definitions handout is discussed. A teacher-made worksheet is used for measuring angles and line segments. PREPARATION REQUIRED NEED SPECIAL MATERIALS
HOUR 7: ____________	**HOUR 8:** ____________	**HOUR 9:** ____________
Point location using angles and distances is demonstrated. Students practice point locations. PREPARATION REQUIRED NEED SPECIAL MATERIALS	Point location using angles and intersecting lines is demonstrated. Students practice point locations. PREPARATION REQUIRED NEED SPECIAL MATERIALS	Students construct drawings by following instructions from a "Point Location Worksheet." ANSWER SHEET PROVIDED HANDOUT PROVIDED NEED SPECIAL MATERIALS

PROJECT CALENDAR:

HOUR 10: ____________ Students create their own point location problems and write out instructions. NEED SPECIAL MATERIALS	**HOUR 11:** ____________ Review for test. NEED SPECIAL MATERIALS	**HOUR 12:** ____________ Final test. TEST PROVIDED NEED SPECIAL MATERIALS
HOUR 13: ____________ Graded tests are returned, followed by a discussion of why geometry and point location, along with the use of drawing equipment, are important. RETURN STUDENT WORK	**HOUR 14:** ____________	**HOUR 15:** ____________
HOUR 16: ____________	**HOUR 17:** ____________	**HOUR 18:** ____________

Lesson Plans and Notes

Hours 1, 2, and 3

Introduction to: Parallel lines
Bisecting line segments
Bisecting angles

HOUR 1: Distribute the handout of terms and definitions that will be used throughout the project. Introduce students to the equipment and tell them to draw examples of line segments and circles. After drawing examples on paper at their desks, they are to place points at both ends of each line segment and at the center of each circle they have drawn. These points are labeled with letters. Students then accurately measure the length of their line segments and the radii of their circles. You may find that much of the hour is needed to explain the mechanics of drawing a circle with a compass. Be sure to demonstrate that the distance between the point of a compass and the tip of the pencil is equal to the *radius* of the circle that can be drawn.

Give students a simple definition of parallel lines (two lines in the same plane that never touch; or, two lines that are always the same distance apart). Have students draw several sets of parallel lines by drawing one line and then carefully measuring from both ends to create a parallel line. This skill is useful for making lettering guides and border lines, for centering pictures or drawings, and for producing neat, presentable posters and other display materials. Tell students that after this course they will be expected to produce posterwork that is neat and properly laid out.

Note:

- For this project to work smoothly, a protractor, compass, and ruler are needed for every student every hour, except the last (Hour 13).

HOUR 2: Using examples on the board, teach students how to use a compass and a straightedge to bisect line segments and angles. Spend the hour explaining and having students practice these skills. It is a good idea to supply a handout with a variety of line segments and angles to be bisected.

HOUR 3: Give students a worksheet assignment to work on during the hour. *This worksheet is to be made by you.* It should consist of a variety of line segments and angles to be bisected. Be sure to provide plenty of room for each problem.

Examples:

- How to Bisect Angles

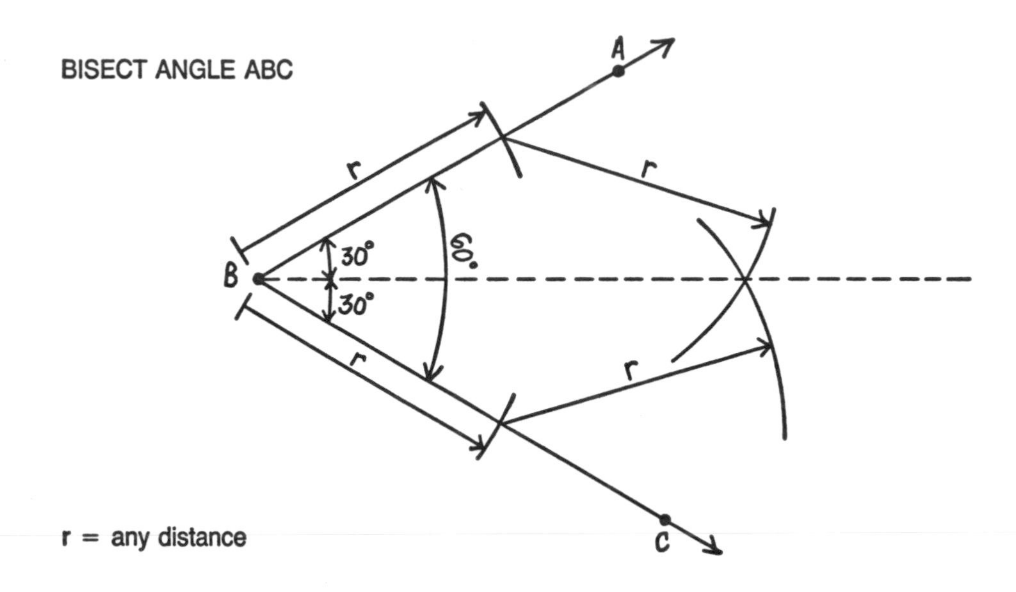

- How to Bisect Line Segments

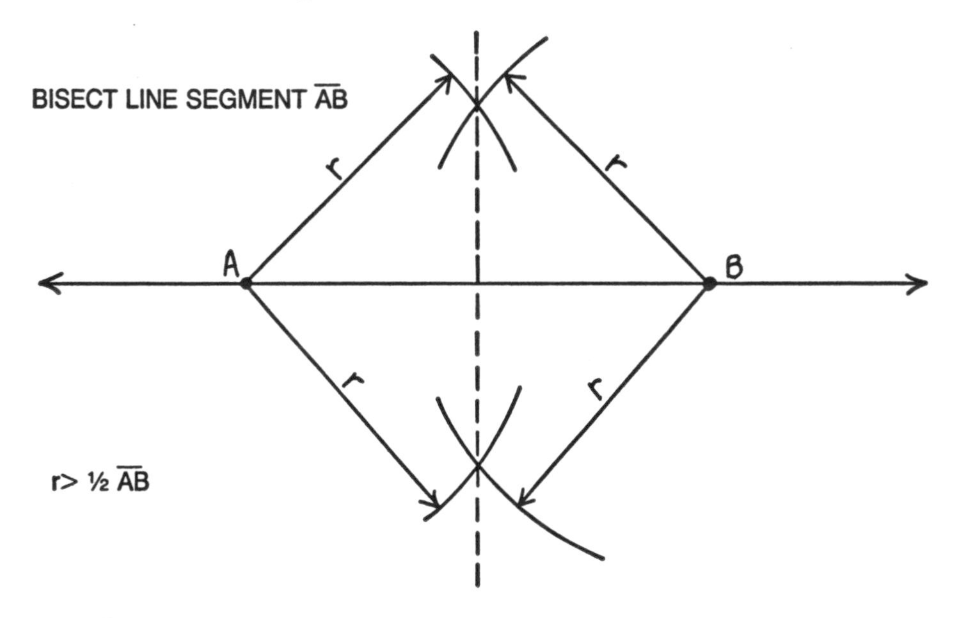

Hours 4 and 5

Circumscribing and Inscribing Triangles with Circles

HOUR 4: Show students how to circumscribe a triangle with a circle; present a step-by-step demonstration on the board (illustrations are provided that show

how to do this). Students then work on circumscribing triangles at their desks, working from handouts that you have made. These handouts should be examples of simple isosceles or equilateral triangles large enough for students to work with. Once these simple triangles are conquered, students should be allowed to make their own, more challenging, triangles to circumscribe.

Notes:

- *Be sure to make the handout triangles big enough to work with.* Also, place heavy emphasis on the importance of neatness and accuracy. Grade the construction lines, not the finished circle. When students have conquered the simple triangles on your worksheet, let them create more difficult triangles to work with on their own.
- The following is an example of how to circumscribe a triangle with a circle:

Circumscribe:

1. Bisect each side.
2. Intersect the bisection lines (inside the triangle here).
3. Use this point as the center of the circle.

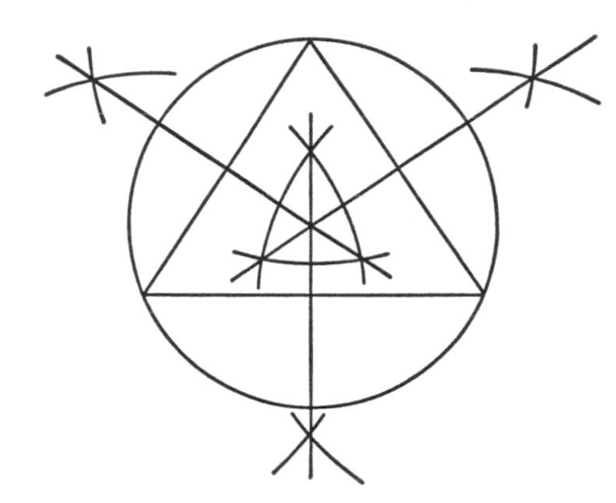

HOUR 5: Show students how to inscribe a circle within a triangle. They then work on inscribing at their desks, using triangle worksheets that are given to them. (They can use new copies of the same worksheet that was given out for Hour 4.)

Note:

- The following is an example of how to inscribe a triangle with a circle:

Inscribe:

1. Bisect each angle.
2. Intersect the bisection lines (always inside the triangle).
3. Use this point as the center of the circle.

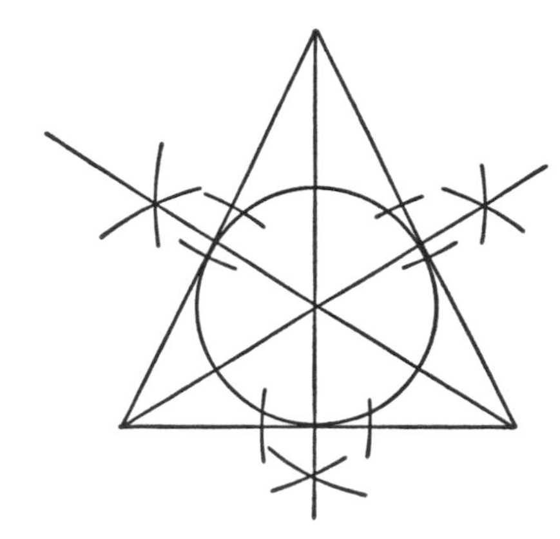

Hours 6, 7, 8, 9, and 10

Point Locations

HOUR 6: Teach students how to measure angles with a protractor and how to accurately measure distance in inches and centimeters with a ruler. If many students have difficulty with this, spend another hour on it. Vocabulary words from the "Protractor–Ruler–Compass: Terms and Definitions" handout are discussed.

1. vertex
2. ray
3. acute angle
4. obtuse angle
5. bisect
6. arc
7. intersect
8. line segment

Distribute a teacher-made worksheet with a variety of angles and line segments to be measured. Require students to record the measurement of each angle in degrees and the length of each line segment in inches *and* centimeters. Students who catch on to this quickly should be allowed to draw additional angles and line segments for one another to measure, while students who have difficulty measuring angles and lines should be given further help. It might be helpful to assign homework for students who are slow to understand.

HOUR 7: Show students how to use angles and distances to locate points on a piece of paper. Demonstrate this on the board, using a line with two points, "A" and "B," on it. A third point, "C," is also on the board, but not on the line that contains segment $\overline{AB}$. By connecting "A" and "C," students see an angle created. This angle can be measured, and the distance from "A" to "C" can also be measured. This information locates point "C" exactly (in the plane of the chalkboard). The angle thus formed is called angle BAC. *The middle letter is always the vertex of the angle!*

At their desks students follow precise directions to put a base line along the long edge of their drawing paper. Instruct the students: "Place your paper horizontally in front of you on the desk, so that the long edges are at the top and bottom. Draw a line across your paper $1\frac{1}{2}$ inches from, and parallel to, the top. Place a point on this line two inches from the left edge and label it point "A." Place a second point on the line seven inches to the right of point "A." Label this point "B." Now locate point "C." It is on a line that makes angle BAC = 30°. It is $3\frac{3}{4}$ inches from point "A." When you have found point "C" raise your hand."

When everyone has found point "C," identify a few more points (that is, point "D": angle BAD = 45°; $\overline{AD}$ = $4\frac{1}{8}$ inches). By the end of this hour students should be comfortable with their ability to find points using angles and distances from two known points on a base line. If they are not, practice in class for another hour.

Example:

- How to locate a point from a base line (angle and distance method)

Figure 1

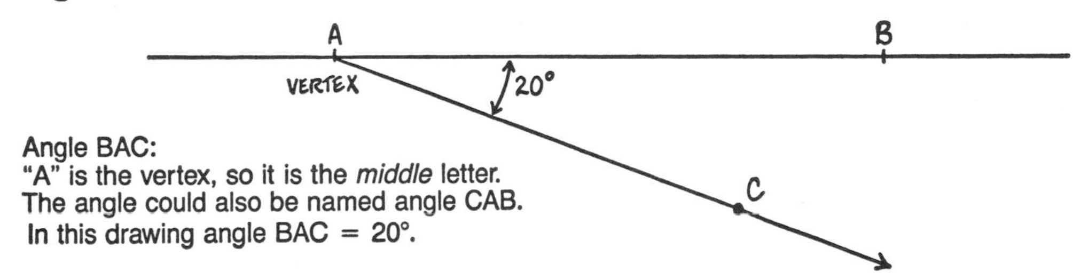

A
B
VERTEX
20°
C
Angle BAC:
"A" is the vertex, so it is the *middle* letter.
The angle could also be named angle CAB.
In this drawing angle BAC = 20°.

Figure 2

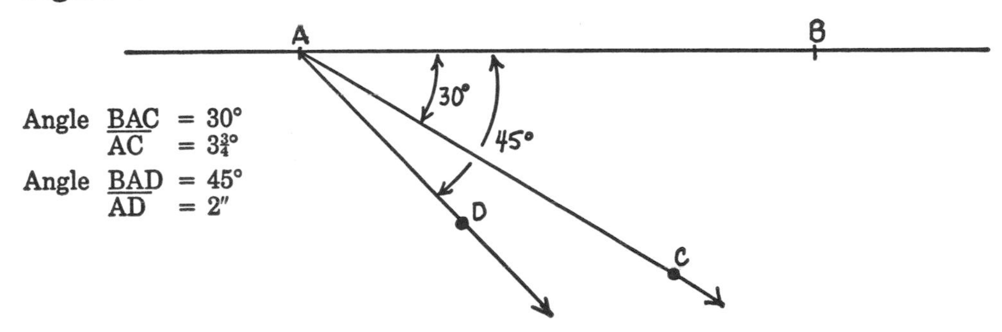

HOUR 8: Students prepare their drawing paper with a base line and points "A" and "B," just as in the previous hour. This time, however, they are shown how point "C" can be located by knowing angle BAC *and* angle ABC. If angle BAC = 29° and angle ABC = 67°, then point "C" can be located at the intersection of line segments $\overline{AC}$ and $\overline{BC}$. When everyone has found point "C," a few more points are identified (that is, point "D": angle BAD = 94°; angle ABD = 31°). By the end of this hour students should be comfortable with their ability to find points using two angles from known points on a base line. Practice in class for another hour if necessary.

Example:

- How to locate a point from a base line (two-angle method)

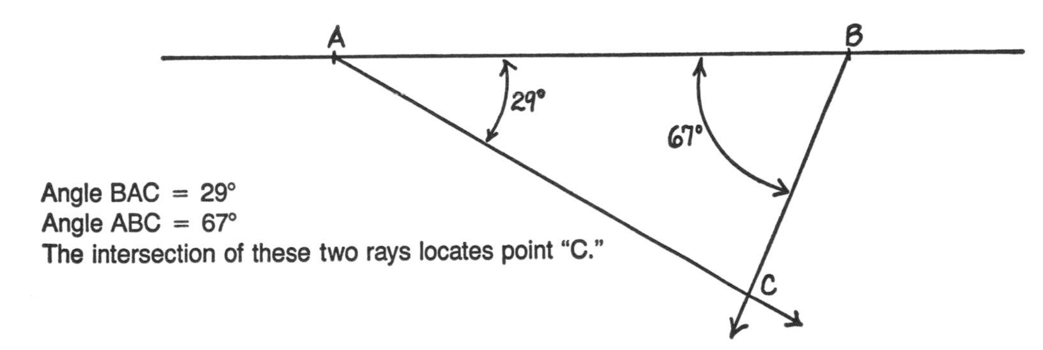

HOUR 9: Give students the "Point Location Worksheet" (provided) and tell them to construct drawings on paper, using instructions that explain how to locate points, and which points to connect. For example, tell students to measure the following angles and distances and then connect points "C," "D," and "E" as they are instructed.

Angle	Distance	Instructions
ABC = 16°	BC = $2\frac{15}{16}$ in	connect: "C" and "D"
ABD = 40°	BD = $6\frac{5}{8}$ in	connect: "D" and "E"
ABE = 78°	BE = $4\frac{3}{8}$ in	connect: "E" and "C"

If these instructions are followed, a triangle will be drawn below the base line.

Since some students will work faster than others, the "Point Location Worksheet" provides problems of varying degrees of difficulty: some are more challenging than others, and these should be assigned to students who understand the concept of point locations. Meanwhile, students who have difficulty can be given individualized help.

HOUR 10: Allow students to create their own point location problems for classmates to solve. This is done by making drawings below a base line with two known points and then carefully measuring the angle and distance (or two angles) to each point to be located. Have the students trade their instructions with other students. This is also time that can be spent with students who are still having difficulty.

Note:

- When locating points by using two angles, the distance between the two points on the base line must be known. If this distance changes, the location of points will change and the shape of the drawing will change correspondingly.

HOUR 11: Review for a final test.

HOUR 12: Final test.

Note:

- Problems 1, 2, and 3 on the final test are based on metric measurement (centimeters). If you do not have access to metric rulers, create new problems to replace these three.

HOUR 13: Return the graded final tests. Spend the hour discussing how point locations and drawing equipment can be used in mathematics, art, graphic design, drafting, mechanical drawing, surveying, and other endeavors.

General Notes About This Project:

- The primary purpose of this project is to teach basic skills for using protractors, compasses, and rulers and to emphasize neatness, accuracy, and patience.

- You will have to prepare worksheets for Hours 2, 3, 4, 5, and 6. They are very simple but they require too much space to be provided here.
    a. The worksheet for Hours 2 and 3 should consist of several pages of line segments and angles for students to bisect. Put only two or three line segments or angles on each page so that students have plenty of room for construction lines.
    b. The worksheet for Hours 4 and 5 should consist of several pages of simple isosceles and equilateral triangles that students circumscribe and inscribe with circles. Put only one triangle on each page so that students have plenty of room for construction lines.
    c. The worksheet for Hour 6 should consist of two or three pages of angles and line segments for students to measure. Place emphasis on *accuracy* when students are working on these problems. The worksheet that was made for Hours 2 and 3 can also be used this hour.
- The dimensions on the point location problems can be changed to metric measurements if you wish to place more emphasis on metrics. Problem 2 of the "Point Location Worksheet" is already done in the metric system.
- The "Point Location Worksheet" in this project is a prerequisite to two projects that are presented later in this book: Tree Mapping and Land Surveying. If you intend to conduct either of these projects, be sure to include point location in your introductory material.

Name ______________________________ Date ______________________________

PROTRACTOR–COMPASS–RULER
Terms and Definitions

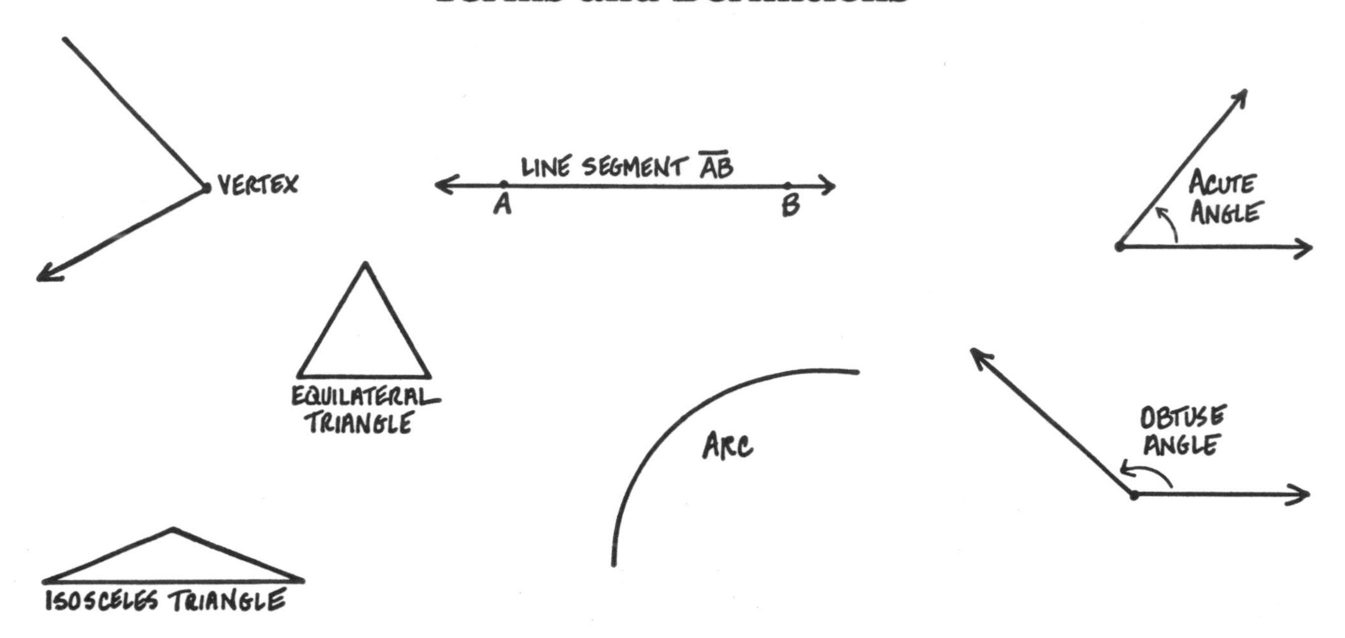

Acute angle: An angle that measures less than 90°.

Angle: The figure formed by two lines extending from the same point. Angles are measured in degrees.

Arc: A portion of a circle drawn with a compass.

Base line: A line between two points, from which other points can be located by measuring angles and distances.

Bisect: To divide an angle or a line segment into two equal parts.

Circumscribe: To draw a circle around any geometric figure so that it touches as many points as possible.

Compass: A tool for drawing circles.

Equilateral triangle: A triangle with all three sides equal in length.

Inscribe: To draw a circle inside any geometric figure so that it touches as many points as possible.

Intersect: To draw two lines so that they cross. The point at which two lines cross is called the intersection of the lines.

Isosceles triangle: A triangle with two equal sides.

Line segment: A portion of a line located between two points. On a line that has points "A" and "B," the segment between "A" and "B" is called line segment AB, which is written $\overline{AB}$.

Obtuse angle: An angle that measures greater than 90° and less than 180°.

Parallel lines: Lines that extend in the same direction and are everywhere the same distance apart.

Protractor: A tool for measuring angles.

Ray: A line that extends in one direction from a point.

Ruler: A tool for measuring distance in inches or centimeters.

Vertex: The point at which two sides of an angle meet.

Name ________________________________ Date ______________________

POINT LOCATION WORKSHEET

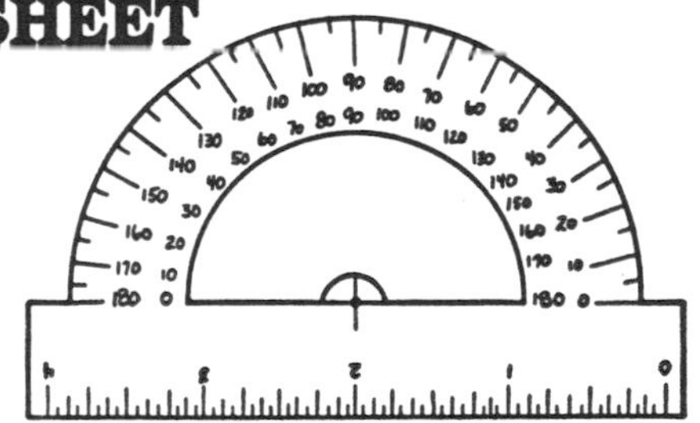

Geometry is perhaps the oldest field of mathematics, aside from addition and subtraction of whole numbers. The ancient Egyptians pioneered the study of geometry out of necessity: the flooding Nile River annually swept away all property markers and they had to be relocated. This could not be done until mathematicians and engineers figured out the relationships among points, lines, angles, and distances. Throughout the ancient world, as people began to live together and claim ownership of land, the study of geometry was of utmost importance for one simple reason: they needed to know how to specifically locate points that could be used as boundary markers. Since then, the discoveries of early students of geometry have been applied to everything from military strategy to space travel. The first major step in understanding geometry is to learn about locating points on a piece of paper. From this knowledge of points, lines, rays, angles, and measurement, comes the foundation for further in-depth and often fascinating studies of geometry.

The point location problems on these handouts should each be done on a separate piece of $8\frac{1}{2}$-by-11-inch drawing paper. Begin each problem by drawing a base line on your paper. Here's how: the base line must be parallel to the long edges of the paper, $1\frac{1}{2}$ inches from *one* of these edges. This base line becomes the top of the drawing, so place your paper on the desk with the base line running across the top. Now measure two inches from the left edge of the paper along the base line and make a point. Mark this point "A." Measure seven inches to the right from point "A" along the base line and make another point. Mark this point "B." This means that $\overline{AB}$ (line segment AB) equals seven inches.

Now that you have a *line segment* ($\overline{AB}$) on the paper, you can make *angles* by drawing *rays* from either point "A" or point "B." Try it on a practice piece of paper: draw a ray from point "A" to create an angle. The *vertex* of the angle is the point you drew the ray from. Now put a point on the ray and mark the point "C." You have created angle BAC; "A" is the middle letter because it is the vertex.

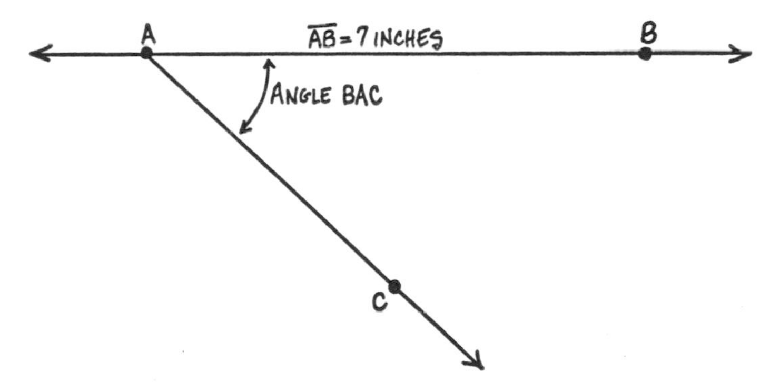

If you know how many degrees are in an angle, it can be drawn accurately on paper. Suppose you are told that "B" is the vertex of a 30° angle. You can draw it:

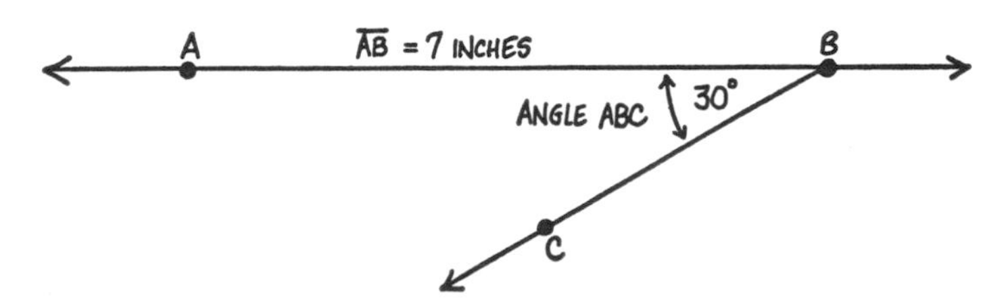

If you put point "C" on the ray, then the angle is ABC, and you can say that angle ABC = 30°.

Name ________________________________ Date ____________________

POINT LOCATION WORKSHEET (continued)

Now, if you know how far "C" is from "B" on the ray, and if you know how many degrees are in angle ABC, you can find point "C" precisely. Suppose you are told that angle ABC = 45° and the distance from "B" to "C" is 1½ inches. From the base line you can locate point "C":

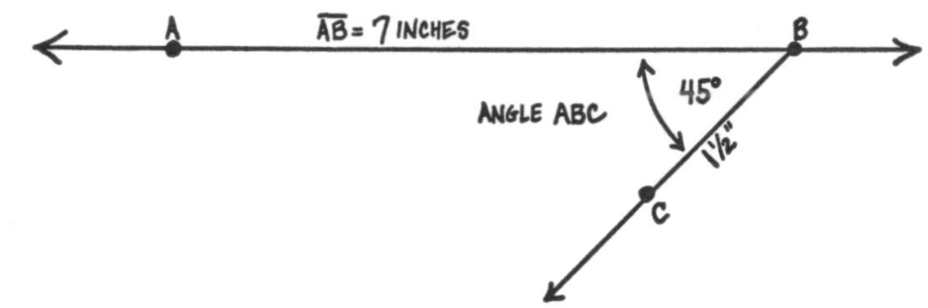

$\overline{BC}$ is a line segment that is 1½ inches long, so the instructions for finding "C" would look like this: angle ABC = 45°; $\overline{BC}$ = 1½ inches.

There is another way to locate point "C." If you know what the angle to "C" is with "A" as the vertex, *and* what the angle to "C" is with "B" as the vertex, then it should look like the following:

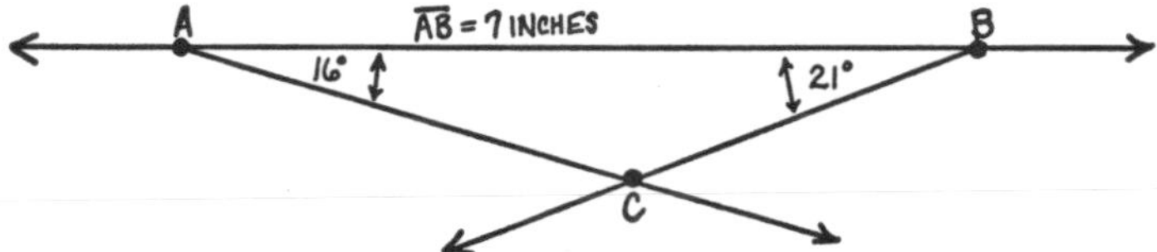

Angle BAC = 16° and angle ABC = 21°. With this information you can accurately find point "C" by intersecting the rays from "A" and "B." (The intersection of two lines makes a point.)

Using this basic review of information about point location, solve the five problems below. *REMEMBER!* The *middle* letter is the vertex of an angle.

PROBLEM 1

Angle	Distance
ABC = 69°	$\overline{BC}$ = $\frac{7}{8}$ in.
ABD = 104°	$\overline{BD}$ = $2\frac{15}{16}$ in.
ABE = 47°	$\overline{BE}$ = $9\frac{1}{16}$ in.
ABF = 33°	$\overline{BF}$ = $8\frac{5}{8}$ in.

Instructions: Connect points C and D, C and F, F and E, and E and D.

PROBLEM 2

Angle	Distance
BAC = 12°	$\overline{AC}$ = 14.4 cm
BAD = 29°	$\overline{AD}$ = 10.5 cm
BAE = 37°	$\overline{AE}$ = 5.1 cm
BAF = 43°	$\overline{AF}$ = 16.4 cm
BAG = 43°	$\overline{AG}$ = 19.2 cm
BAH = 61°	$\overline{AH}$ = 12.7 cm
BAI = 73°	$\overline{AI}$ = 13.7 cm

Instructions: Connect points C and E, E and I, I and G, G and C, D and H, H and F, and F and D.

PROBLEM 3

Angles with "A" as Vertex	Angles with "B" as Vertex
BAC = 11°	ABC = 27°
BAD = 27°	ABD = 97°
BAE = 53°	ABE = 73°
BAF = 73°	ABF = 53°
BAG = 97°	ABG = 27°
BAH = 27°	ABH = 11°

Instructions: Connect points C and D, D and E, E and F, F and G, G and H, and H and C.

Name ______________________________ Date ________________________

POINT LOCATION WORKSHEET (continued)

PROBLEM 4

Angle	Distance
BAC = 16°	$\overline{AC} = 4\frac{1}{4}$ in.
BAD = 36°	$\overline{AD} = 7\frac{1}{8}$ in.
BAE = 38°	$\overline{AE} = 6\frac{7}{8}$ in.
BAF = 44°	$\overline{AF} = 6\frac{3}{4}$ in.
BAG = 50°	$\overline{AG} = 8$ in.
BAH = 52°	$\overline{AH} = 8\frac{3}{4}$ in.
BAI = 54°	$\overline{AI} = 6\frac{3}{4}$ in.
BAJ = 55°	$\overline{AJ} = 5\frac{5}{8}$ in.
BAK = 58°	$\overline{AK} = 4\frac{7}{8}$ in.
BAL = 61°	$\overline{AL} = 4\frac{3}{4}$ in.
BAM = 64°	$\overline{AM} = 6\frac{7}{8}$ in.
BAN = 69°	$\overline{AN} = 7\frac{3}{8}$ in.
BAO = 31°	$\overline{AO} = 8\frac{1}{8}$ in.
BAP = 75°	$\overline{AP} = 4\frac{1}{4}$ in.

Instructions:

1. Connect points C and L, C and D, D and E, E and K, D and O, L and P, K and N, E and H, H and N, L and K.
2. Draw a small circle around point J.
3. Draw the same size circle around point F.
4. Draw a small triangle around point I.
5. Draw a curve from point M to point G, using the curved edge of your protractor.

PROBLEM 5

Angles with "A" as Vertex	Angles with "B" as Vertex
BAC = 15°	ABC = 28°
BAD = 20°	ABD = 21°
BAE = 70°	ABE = 25°
BAF = 54°	ABF = 30°
BAG = 30°	ABG = 85°
BAH = 39°	ABH = 58°
BAI = 69°	ABI = 47°
BAJ = 52°	ABJ = 60°
BAK = 38°	ABK = 87°
BAL = 82°	ABL = 49°
BAM = 59°	ABM = 66°

Instructions: Connect C and D, D and E, E and F, F and C, C and H, F and I, E and L, I and H, J and G, J and M, G and K, M and K, J and I, H and G, L and M.

Name ______________________ Date ______________________

PROTRACTOR–COMPASS–RULER
Final Test

PROBLEM 1

Draw a border around a piece of 8½-by-11-inch drawing paper. Each border line should be two centimeters from the edge of the paper and parallel to it.

PROBLEM 2

With your bordered paper placed horizontally on the desk, place a point on the top line, five centimeters from the left edge of the paper. Label this point "A." Place a second point on the top line eight centimeters from the right edge of the paper. Label this point "B." ($\overline{AB}$ = 15 cm)

PROBLEM 3

Locate the following points on your paper by measuring angles and distances from point "A."

Angle	Distance
BAC = 8°	AC = 9.1 cm
BAD = 27°	AD = 14.5 cm
BAE = 36°	AE = 11.1 cm
BAF = 36°	AF = 22.4 cm
BAG = 46°	AG = 18.5 cm
BAH = 53°	AH = 8.2 cm
BAI = 56°	AI = 16.0 cm
BAJ = 63°	AJ = 19.3 cm
BAK = 69°	AK = 14.1 cm
BAL = 90°	AL = 13.2 cm

Instructions: Connect the following:

C and D
D and F
C and H
H and L
H and E
D and E
C and E
E and I
I and J
H and K
D and G
L and K
K and I
I and G
G and F

Name ______________________________ Date ______________________________

PROTRACTOR–COMPASS–RULER

Final Test (continued)

PROBLEM 4

Inscribe a circle in the triangle below.

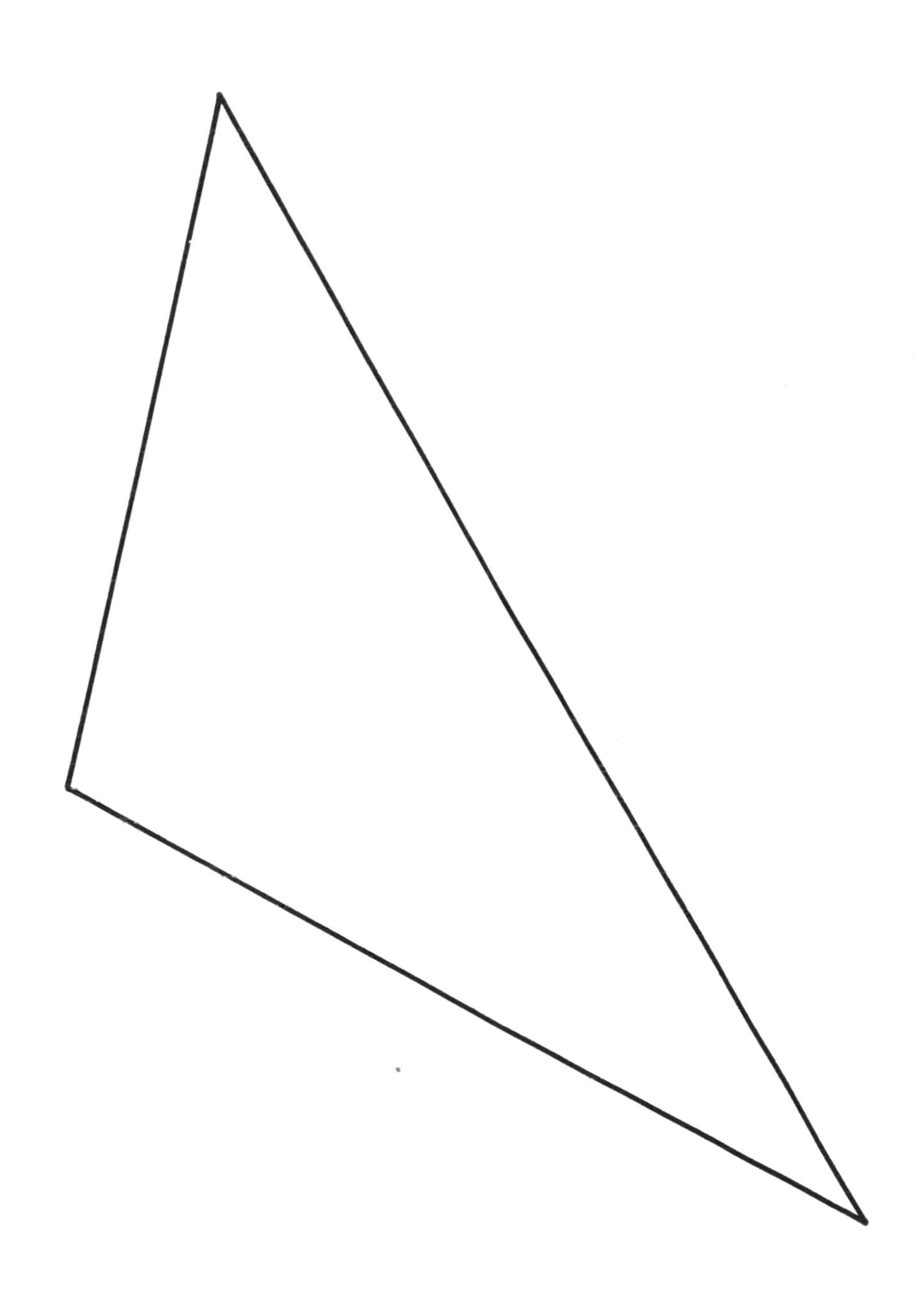

Name ______________________________ Date ______________________________

PROTRACTOR–COMPASS–RULER

Final Test (continued)

PROBLEM 5

Circumscribe the triangle below with a circle.

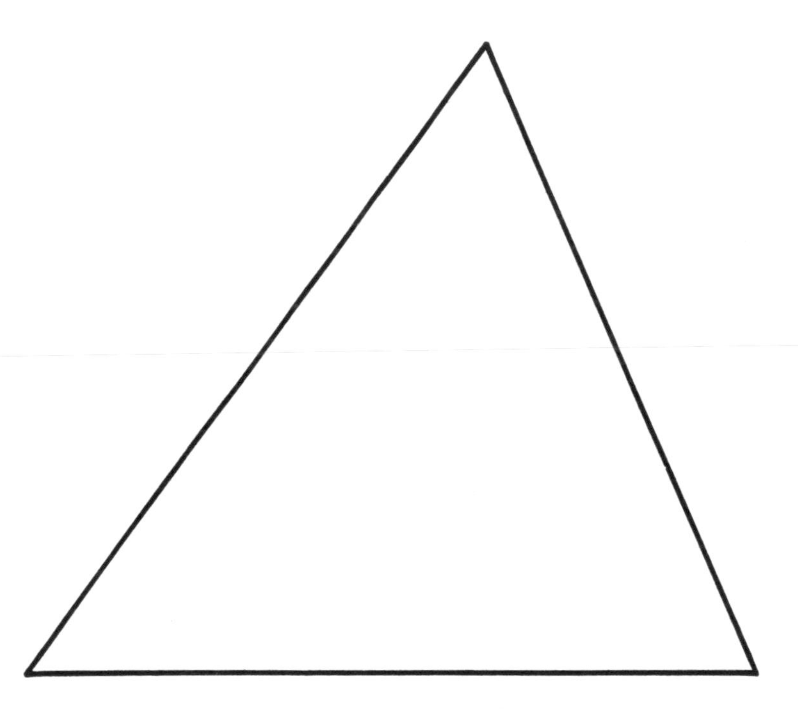

TEACHER ANSWER SHEET: PROBLEM 1
(Point Location Worksheet)

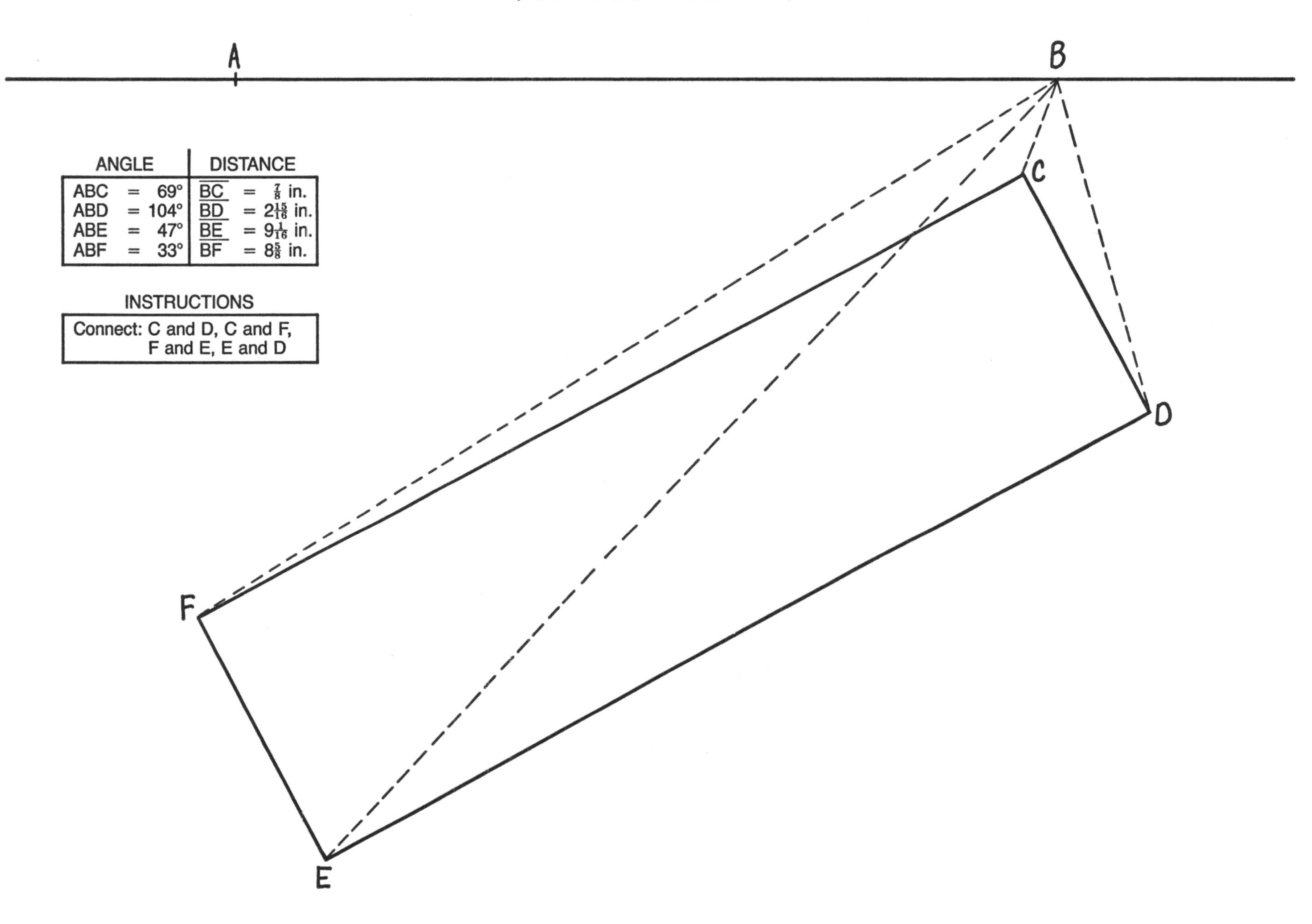

ANGLE	DISTANCE
ABC = 69°	$\overline{BC}$ = $\frac{7}{8}$ in.
ABD = 104°	$\overline{BD}$ = $2\frac{15}{16}$ in.
ABE = 47°	$\overline{BE}$ = $9\frac{1}{16}$ in.
ABF = 33°	$\overline{BF}$ = $8\frac{5}{8}$ in.

INSTRUCTIONS

Connect: C and D, C and F,
F and E, E and D

TEACHER ANSWER SHEET: PROBLEM 2
(Point Location Worksheet)

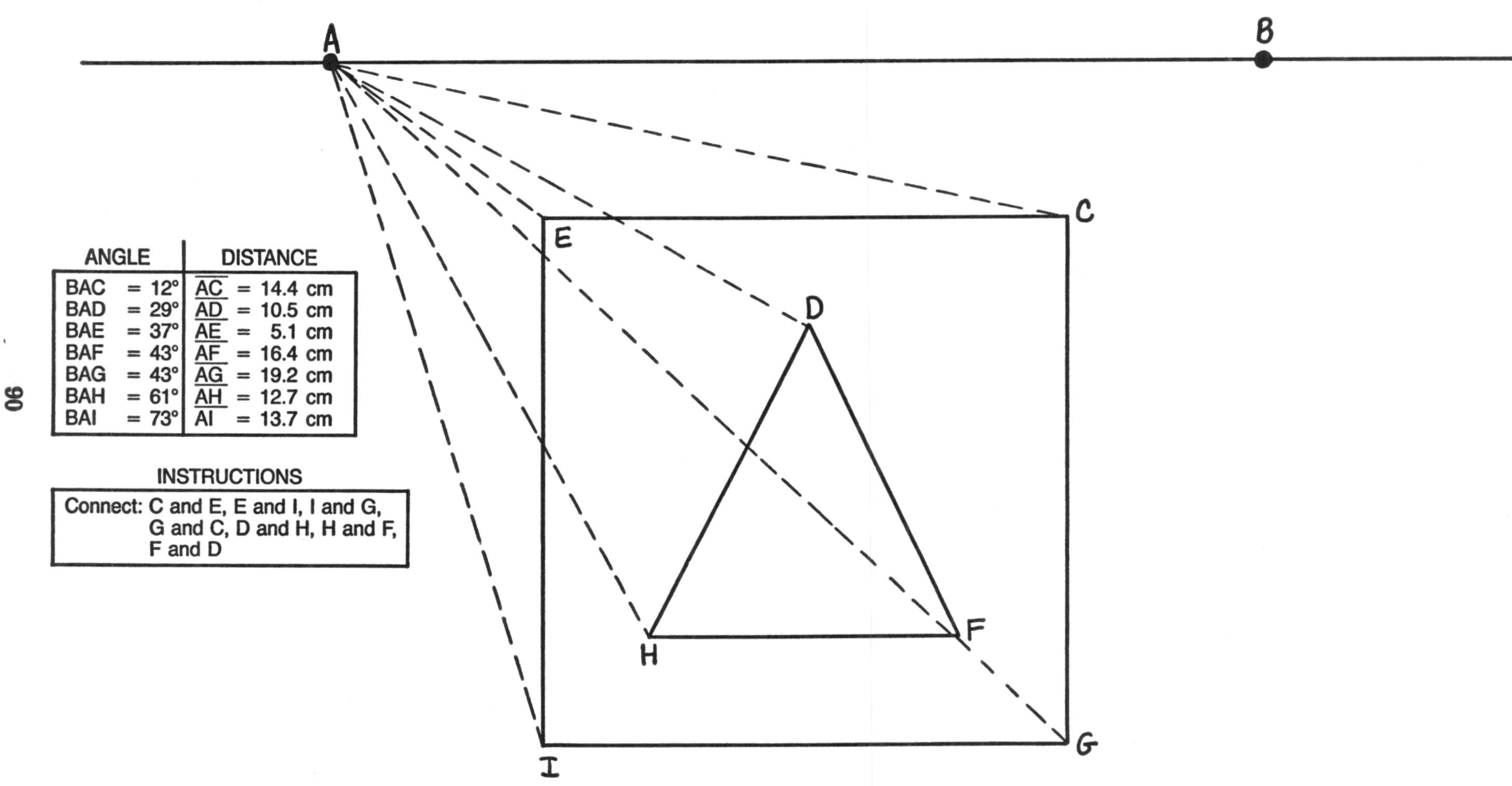

ANGLE	DISTANCE
BAC = 12°	$\overline{AC}$ = 14.4 cm
BAD = 29°	$\overline{AD}$ = 10.5 cm
BAE = 37°	$\overline{AE}$ = 5.1 cm
BAF = 43°	$\overline{AF}$ = 16.4 cm
BAG = 43°	$\overline{AG}$ = 19.2 cm
BAH = 61°	$\overline{AH}$ = 12.7 cm
BAI = 73°	$\overline{AI}$ = 13.7 cm

INSTRUCTIONS

Connect: C and E, E and I, I and G, G and C, D and H, H and F, F and D

TEACHER ANSWER SHEET: PROBLEM 3
(Point Location Worksheet)

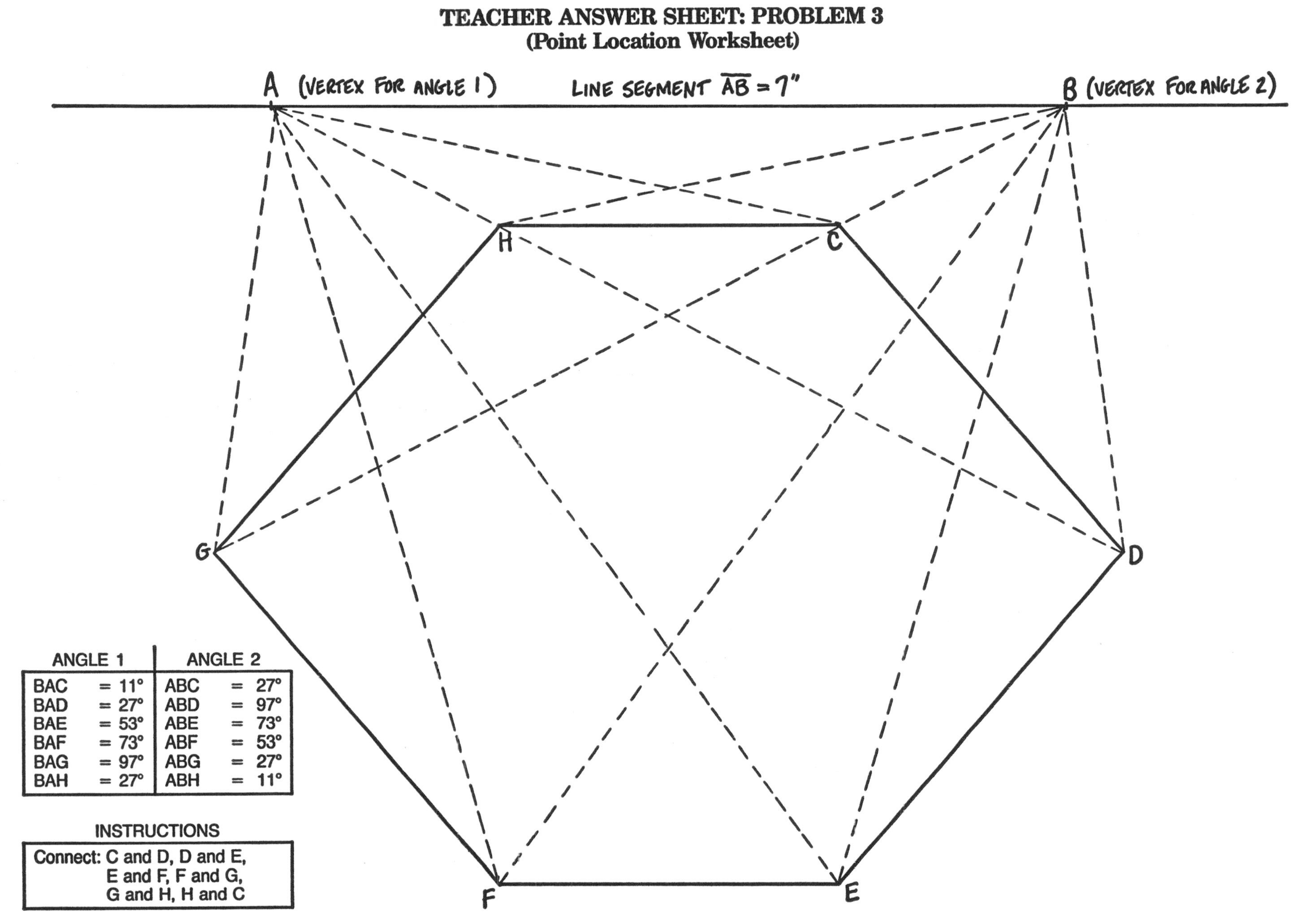

ANGLE 1		ANGLE 2	
BAC	= 11°	ABC	= 27°
BAD	= 27°	ABD	= 97°
BAE	= 53°	ABE	= 73°
BAF	= 73°	ABF	= 53°
BAG	= 97°	ABG	= 27°
BAH	= 27°	ABH	= 11°

INSTRUCTIONS

Connect: C and D, D and E,
E and F, F and G,
G and H, H and C

TEACHER ANSWER SHEET: PROBLEM 4
(Point Location Worksheet)

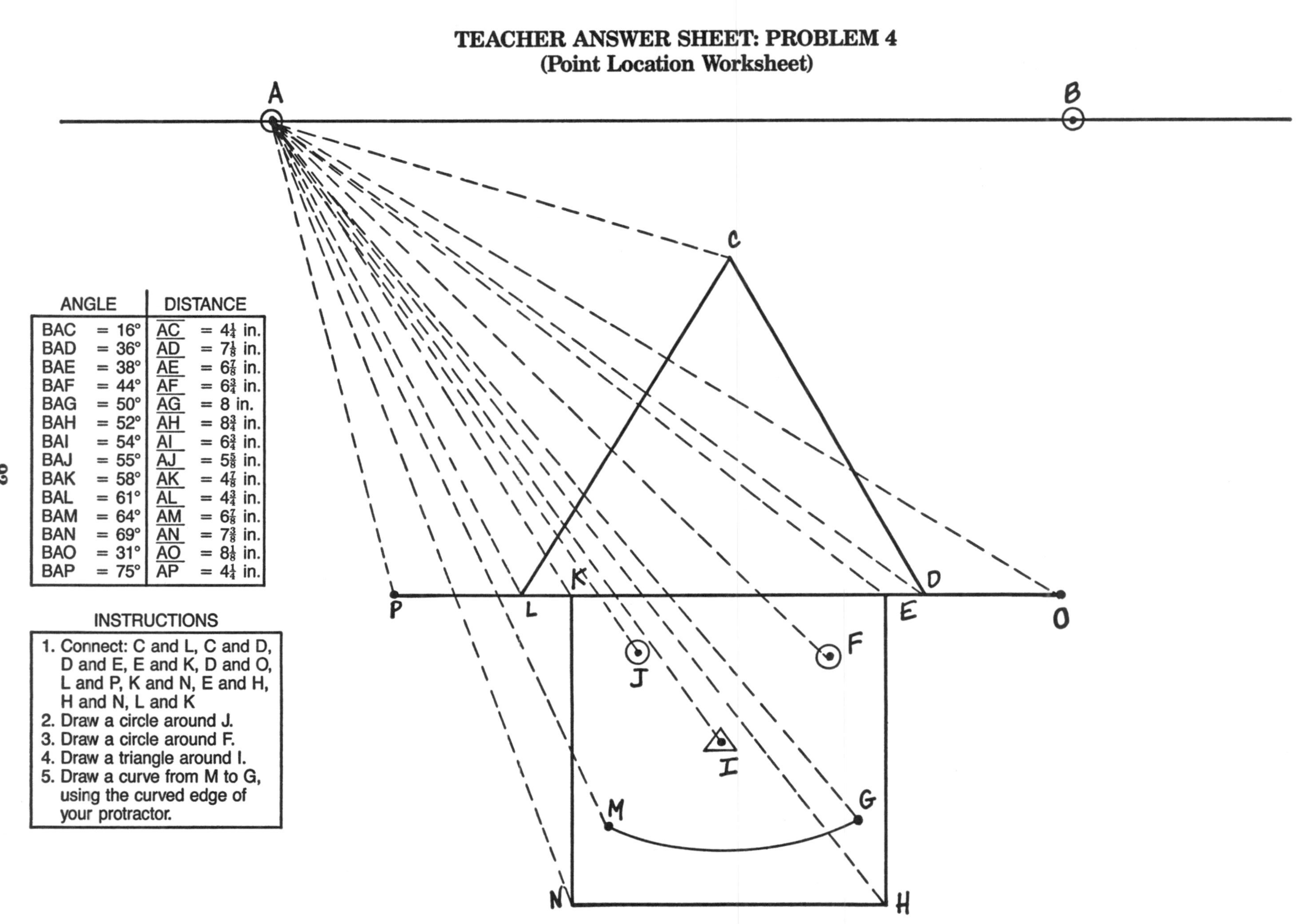

ANGLE	DISTANCE
BAC = 16°	AC = $4\frac{1}{4}$ in.
BAD = 36°	AD = $7\frac{1}{8}$ in.
BAE = 38°	AE = $6\frac{7}{8}$ in.
BAF = 44°	AF = $6\frac{3}{4}$ in.
BAG = 50°	AG = 8 in.
BAH = 52°	AH = $8\frac{3}{4}$ in.
BAI = 54°	AI = $6\frac{3}{4}$ in.
BAJ = 55°	AJ = $5\frac{5}{8}$ in.
BAK = 58°	AK = $4\frac{7}{8}$ in.
BAL = 61°	AL = $4\frac{3}{4}$ in.
BAM = 64°	AM = $6\frac{7}{8}$ in.
BAN = 69°	AN = $7\frac{3}{8}$ in.
BAO = 31°	AO = $8\frac{1}{8}$ in.
BAP = 75°	AP = $4\frac{1}{4}$ in.

INSTRUCTIONS

1. Connect: C and L, C and D, D and E, E and K, D and O, L and P, K and N, E and H, H and N, L and K
2. Draw a circle around J.
3. Draw a circle around F.
4. Draw a triangle around I.
5. Draw a curve from M to G, using the curved edge of your protractor.

TEACHER ANSWER SHEET: PROBLEM 5
(Point Location Worksheet)

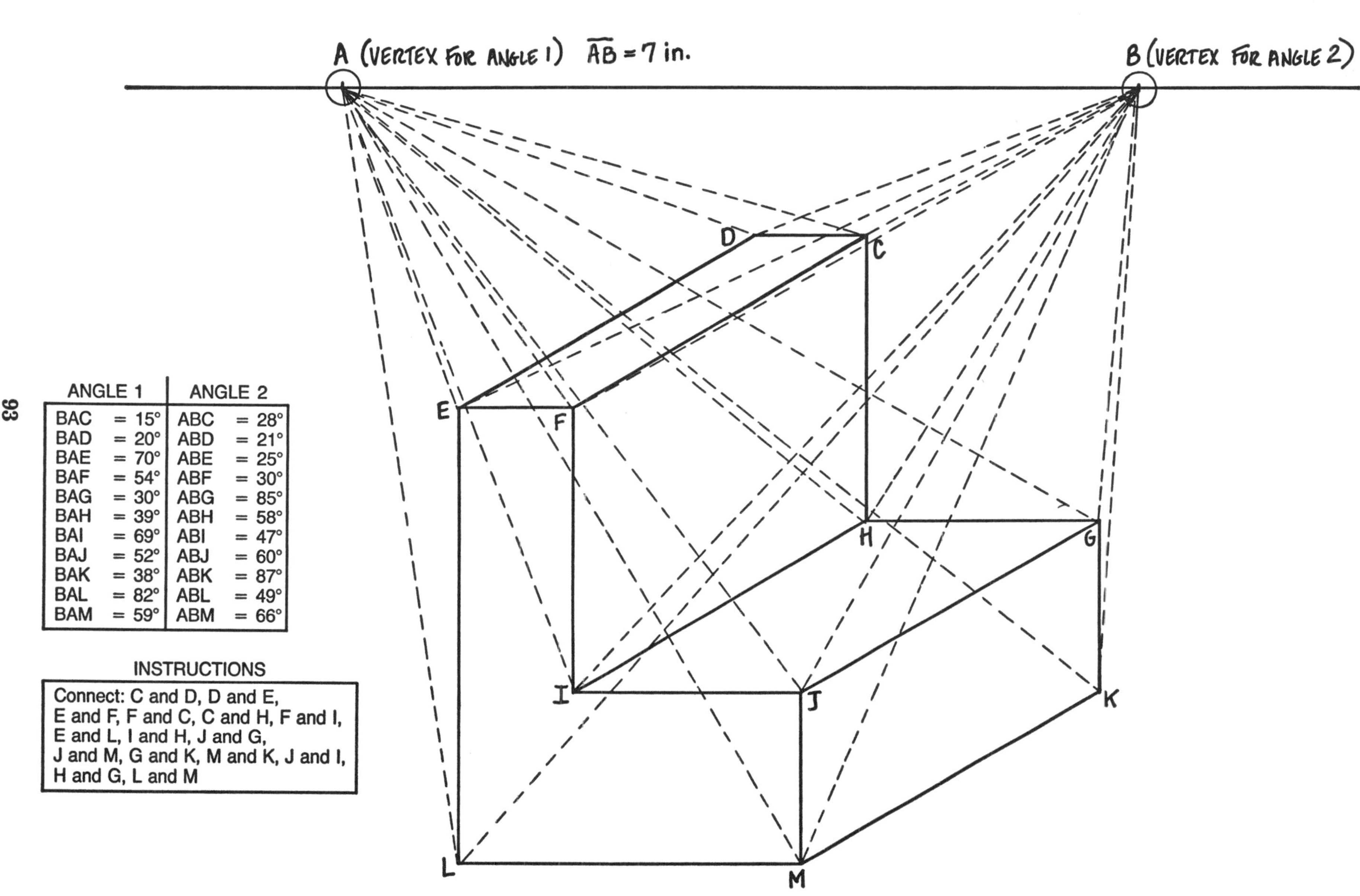

ANGLE 1		ANGLE 2	
BAC	= 15°	ABC	= 28°
BAD	= 20°	ABD	= 21°
BAE	= 70°	ABE	= 25°
BAF	= 54°	ABF	= 30°
BAG	= 30°	ABG	= 85°
BAH	= 39°	ABH	= 58°
BAI	= 69°	ABI	= 47°
BAJ	= 52°	ABJ	= 60°
BAK	= 38°	ABK	= 87°
BAL	= 82°	ABL	= 49°
BAM	= 59°	ABM	= 66°

INSTRUCTIONS

Connect: C and D, D and E, E and F, F and C, C and H, F and I, E and L, I and H, J and G, J and M, G and K, M and K, J and I, H and G, L and M

TEACHER ANSWER SHEET:
Final Test Problem 3

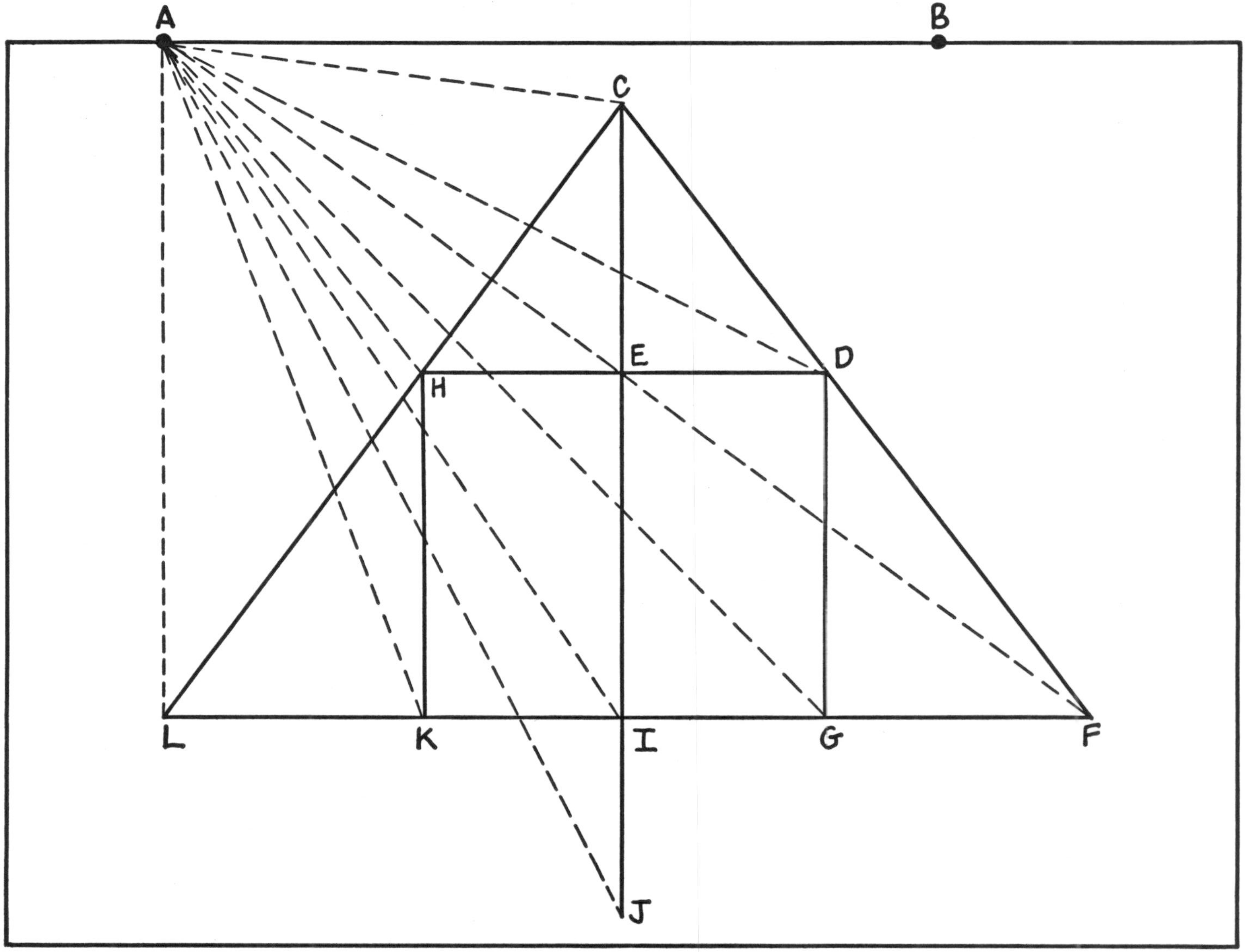

A
B
C
D
E
H
L
K
I
G
F
J

M-6

TREE MAPPING

Teacher Preview

General Explanation:

Using the concept of point location presented in the preceding project (Protractor–Compass–Ruler), this project emphasizes mapping an area by accurately locating some of the trees (or other physical objects) from an established base line. The project can be modified and conducted indoors if an outdoor project is not practical or possible. Some method of measuring angles is necessary: plans for a surveyor's transit are included, along with some suggestions for simpler methods. This project is designed on the assumption that homemade transits will be used.

Length of Project: 11 hours

Level of Independence: Intermediate

Goals:

1. To make use of point location skills learned in previous courses and projects.
2. To provide an activity that emphasizes geometry, data collection, scale drawing, and other mathematics skills.
3. To allow students to work independently to create maps that show the location of trees within a specified area.
4. To take students outside for an educational activity.
5. To provide an opportunity to teach students about trees (optional).

During This Project Students Will:

1. Measure angles and distances from points on a base line.
2. Record data as field notes.
3. Make accurate scale drawings from their field data.
4. Identify the trees they locate (optional).

Skills:

Using a compass
Using a protractor
Using a ruler
Using a straightedge
Linear measurement

Accepting responsibility
Concentration
Controlling behavior
Individualized study habits
Persistence

Angular measurement
Collecting data
Neatness
Accuracy
Sense of "quality"
Listening
Organizing
Sharing space
Taking care of materials
Drawing/sketching/graphing
Scale drawing
Point location
Observing
Working in groups

Handouts Provided:

- "Practice Field Notes"
- "Field Notes Example"
- "Tree Map Example"
- "Tree Mapping Student Assignment Sheet"

Additional Material Provided:

- "Transit Materials"
- "Transit Construction Diagram"

PROJECT CALENDAR:

HOUR 1: ______	HOUR 2: ______	HOUR 3: ______
Point location skills, using angles and distances from an established base line, are reviewed. PREPARATION REQUIRED NEED SPECIAL MATERIALS	A method for taking field notes is introduced. Students record practice field notes and then are given handouts to check their work. HANDOUTS PROVIDED	Students are taught how to use homemade surveyors' transits. PREPARATION REQUIRED NEED SPECIAL MATERIALS
HOUR 4: ______	**HOUR 5:** ______	**HOUR 6:** ______
The tree-mapping project is explained and students receive assignment sheets. Guidelines for conducting an outdoor project are discussed. HANDOUT PROVIDED	Students go outside to begin the tree-mapping project. Most of the hour is spent explaining the project once again. PREPARATION REQUIRED NEED SPECIAL MATERIALS	Tree mapping continues; students carefully record all data. NEED SPECIAL MATERIALS
HOUR 7: ______	**HOUR 8:** ______	**HOUR 9:** ______
Data collection is completed by the end of the hour. NEED SPECIAL MATERIALS	In the classroom, students work on their maps. NEED SPECIAL MATERIALS	Students continue to work on their maps. NEED SPECIAL MATERIALS

PROJECT CALENDAR:

HOUR 10: ______	HOUR 11: ______	HOUR 12: ______
Maps are finished by the end of the hour and are turned in, along with field notes. NEED SPECIAL MATERIALS STUDENTS TURN IN WORK	Graded maps and field notes are returned, followed by a discussion of the project. RETURN STUDENT WORK	
HOUR 13: ______	**HOUR 14:** ______	**HOUR 15:** ______
HOUR 16: ______	**HOUR 17:** ______	**HOUR 18:** ______

Lesson Plans and Notes

HOUR 1: Give students a general review of point location techniques, using angles and distances from an established base line. It may be helpful to devise a review quiz for this hour, or provide a handout taken from the Protractor–Compass–Ruler point location project. Protractors and rulers are needed this hour.

HOUR 2: Distribute the "Practice Field Notes" handout at the beginning of the hour. Show students how their field notes are to be kept, and spend the hour recording practice field notes. Use the data from the "Field Notes Example" handout that is provided with this project to give information orally to the class. First, make a drawing on the board that illustrates a base line and trees (exactly like the "Tree Map Example" handout). Then present data verbally to students:

> "The tree lettered D on the board is a 24-inch red maple. The transit is set over point A, and angle BAD = 60°. The transit was turned to the left of the base line to measure this angle. The tree is 93 feet from the transit. (That is to say, 93 feet from point A.)"

Instruct students to record their notes neatly, in columns, and to use this same method when they go outside to do field work. Give students the "Field Notes Example" and the "Tree Map Example" handouts at the *end* of the hour so they can check their work and use it for later reference. Practice field notes may be collected and graded.

Note:

- When the transit is set over point "A" and is sighted on point "B," an angle can be turned either to the right or to the left. If a tree is 30° to the left of "B," this fact must be recorded. Otherwise, when it comes time to draw a map from notes, students won't know if angle BAC, which equals 30°, lies to the left or right of "B" as viewed from "A." Also, the "size" of a tree is its *diameter* in inches. This should be estimated, not calculated.

HOUR 3: Teach students how to measure angles outside using homemade transits. These are the points of emphasis:

1. The transit should be set level, square to the earth.
2. The plumb bob must be over the *vertex* point of any angle that is to be measured. This requires some manipulation, because the transit has to be level and at the same time over the vertex.

TOP VIEW OF THE TRANSIT

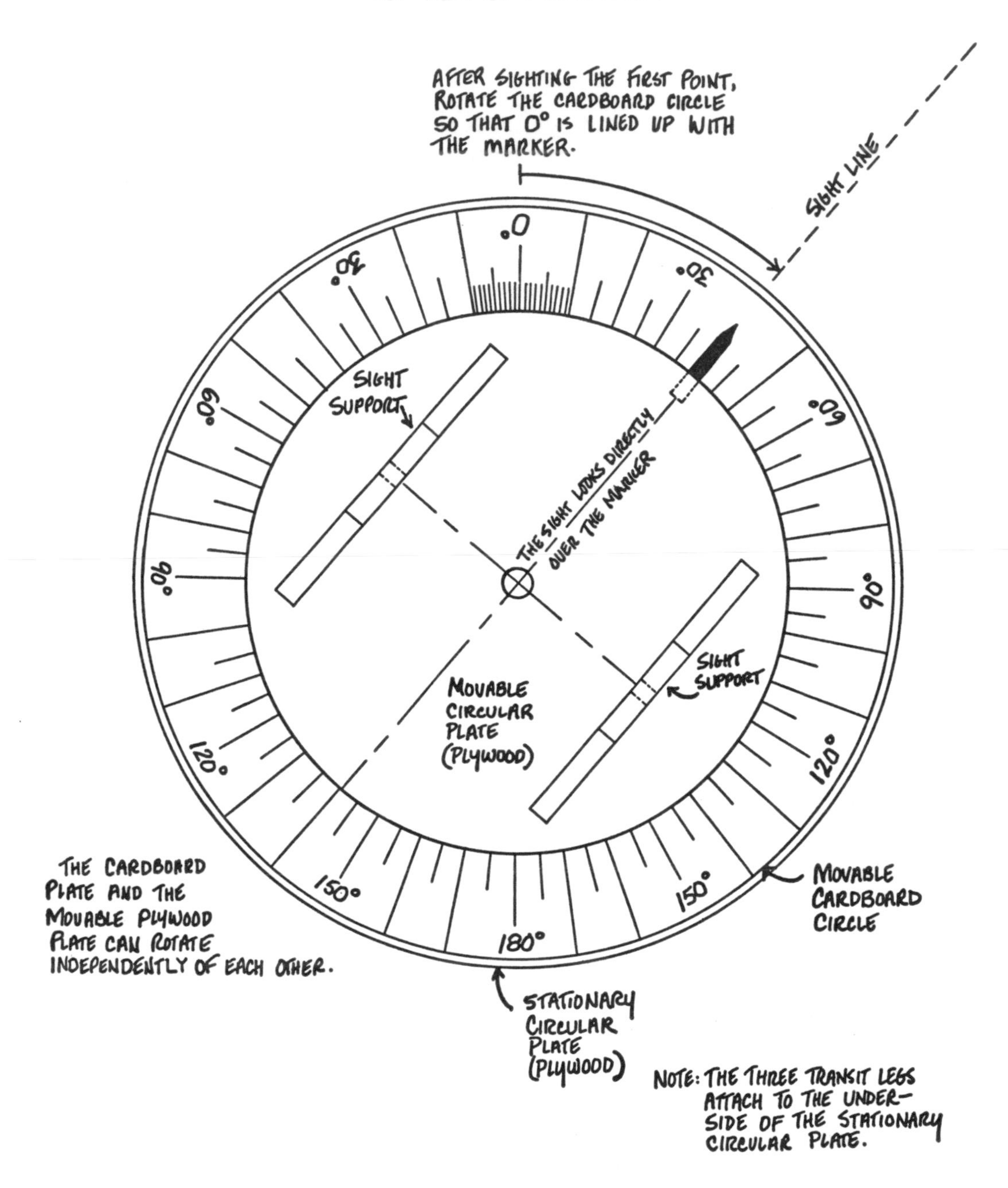
AFTER SIGHTING THE FIRST POINT, ROTATE THE CARDBOARD CIRCLE SO THAT 0° IS LINED UP WITH THE MARKER.
SIGHT LINE
0°
30°
30°
60°
60°
90°
90°
120°
120°
150°
150°
180°
SIGHT SUPPORT
THE SIGHT LOOKS DIRECTLY OVER THE MARKER
SIGHT SUPPORT
MOVABLE CIRCULAR PLATE (PLYWOOD)
THE CARDBOARD PLATE AND THE MOVABLE PLYWOOD PLATE CAN ROTATE INDEPENDENTLY OF EACH OTHER.
MOVABLE CARDBOARD CIRCLE
STATIONARY CIRCULAR PLATE (PLYWOOD)
NOTE: THE THREE TRANSIT LEGS ATTACH TO THE UNDER-SIDE OF THE STATIONARY CIRCULAR PLATE.

3. When starting to measure an angle, aim the transit at the first point to be sighted. This is possible because the sight is mounted on a circular plate that can be rotated 360°. When the point is sighted, move the cardboard circle until 0° is lined up with the marker. Now the transit sight is "zeroed in" on the first point.

4. When turning the transit to sight the second point, the circular cardboard plate is held stationary.
5. After sighting the second point, the angle is read directly from where the marker points on the cardboard plate. If an angle of 60° has been turned, the marker will have stopped at "60°." It is important to note if the 60° is turned to the *left* or to the *right;* this information should be recorded.

Notes:

- If transits are to be built, it must be done before Hour 3. Instructions for constructing a simple, yet functional, surveyor's transit are provided, following the student assignment sheet, at the end of the project. You may be able to make arrangements with an industrial arts class to build several transits for you if you provide the plans and the materials. Or, you may want to have students build their own transits, either at home, at school, or perhaps during a Saturday workshop with parents. Simpler methods of roughly measuring angles can be devised if building transits is not feasible. (See the fifth note below.).
- A plumb bob hung from the center of the transit is necessary to accurately set up over a point. You may want to replace the center bolt that is called for in the transit plans with an eye bolt, with the "eye" pointing down. The plumb line can be tied to the eye.
- For more accuracy, drill small holes in the sight of the transit and stretch crosshairs through the holes. Obviously, one hair should be vertical and one horizontal. Actually, vertical hairs are all that is necessary for point locations.
- It is very important that you clearly understand how to use the surveyor's transit before attempting to teach students. Take it out and work with it until you feel comfortable setting it up over a point, turning an angle and recording the measurement of the angle. Remember, the plumb bob should be over the point and the *transit must be level.* It is not necessary to put a bubble level on the transit, but look at it closely after it is set to be sure it is fairly level and square with the earth.
- Transits are not necessary for this project to be successfully conducted, since emphasis is placed on data collection and mapping rather than on transit skills. Here are two simpler suggestions for determining angles:
  a. Glue or fasten a large protractor to the seat of a sturdy three-legged stool; a short one (one to two feet high) works well. Set the stool over one of the points on the base line, so that the middle of the protractor is above the point, and the 0° mark is aligned with the other point. Stretch a piece of kite string from the center of the protractor toward a tree, and read the angle directly off the protractor.
  b. An even simpler method is to hold a protractor over one of the base-line points, line 0° toward the other point, and carefully sight across the

protractor to a tree to determine the angle. This is called "eyeballing it." Two people are required: one to hold the protractor, and one to eyeball.

HOUR 4: Give students the "Tree Mapping Student Assignment Sheet" and explain the entire project step-by-step. Discuss project guidelines, rules for outdoor conduct, and the procedures that are to be followed. Tell students to dress for an outdoor activity the next three class hours.

HOUR 5: Divide students into mapping groups (three or four students per group), after which they go outside with transits to begin their tree-mapping project. Place emphasis on neatness, accuracy, and thoroughness, not speed. Nobody is expected to locate ten trees (as the assignment sheet specifies) in one hour; most of the time is spent explaining once again how to set the transit over a point and read an accurate angle to a tree. If every group properly locates two trees and records its field data, the hour can be considered a complete success. It is extremely helpful to have additional adult support for this part of the project. Teacher aides, parent volunteers, or even high school students will help the project run smoothly.

Notes:

- It is necessary to have access to an area with plenty of trees. Establish a base line through this area with marked points at each end ("A" and "B"). There should be an unobstructed line of sight between "A" and "B." The base line has to be set before the beginning of Hour 5.
- To avoid congestion and confusion at points "A" and "B," extend the base line beyond each point and drive a permanent sight rod into the ground at each end. These sight rods should be at least six feet high. This way a transit can be on "A" and a group can still get a good sight from "B" by looking *over* the group on "A" to the sight rod beyond. Or, it may be necessary to establish more than one base line: several parallel base lines can be made that are separated by as little as 10 feet. More than one base line should be used if more than 15 students are involved in the project.
- Distances can be measured with a 50- or 100-foot steel or cloth tape. If measuring tapes are not available, have students pace distances. First, mark off an exact distance on a sidewalk or parking lot (100 feet is convenient) and have each student pace this distance at least three times and then calculate his or her average pace length. From this information, fairly accurate distances from base-line points to trees can be determined. See Mathematics in the Park for a pacing handout.

HOUR 6: Students continue to locate trees from the base line and record field data that will allow them to draw an accurate map of the area.

HOUR 7: Students finish their tree locations. Those who complete their work before the hour is over are allowed to locate other physical features in the area: buildings, fences, power poles, fire hydrants, and so forth. They are encouraged to include these features in their tree maps.

HOUR 8: Students begin work on their maps in class. Even though they worked in small groups, each is still responsible for producing his or her own map and set of field notes from the group's data. Each individual starts with a standard piece of drawing paper. Instruct the students to draw a base line through the *center* of their paper and label points "A" and "B." Establish a scale (such as 1 inch = 40 feet) for the entire class. Allow group members to confer, but the maps are to be individually drawn. Students work on their maps until the end of the hour.

Note:

- A scale of 1 inch = 40 feet will fit a 300-foot base line on an $8\frac{1}{2}$-by-11-inch piece of paper. Other scales can be used if they are more convenient. Students should not locate trees that are more than 200 feet from the base line, because these trees may not fit on their maps. REMEMBER: The base line runs through the *center* of the page for this project.

HOUR 9: Students continue to work on their maps.

HOUR 10: Students finish their maps by the end of the hour. If some students finish early, you may want to let them work on a large-scale tree map on posterboard or a strip of banquet table paper. This map can show all the trees in the area, or at least those that were located during the project. Maps and field notes are handed in at the end of the hour.

HOUR 11: Return the maps and field notes to students and discuss them. Then turn the discussion to the project itself and what students gained by participating in it. Place emphasis on the importance of collecting data and taking accurate notes. Also discuss applications of the mathematics used in the project, such as in land surveying, mapping, and studies of geometry. Emphasize the level of independence involved in this project: students need to know how to learn on their own.

General Notes About This Project:

- It is important to understand that although this project was designed as an outdoor activity, it can be conducted in a classroom, recreation room, or gymnasium with some restructuring.
  a. In its simplest form, tree mapping can be conducted on paper with students sitting at their desks in a classroom. To do so, create a handout that has at least 20 trees drawn on it, and a base line drawn through the *center* of the page. *Example:*

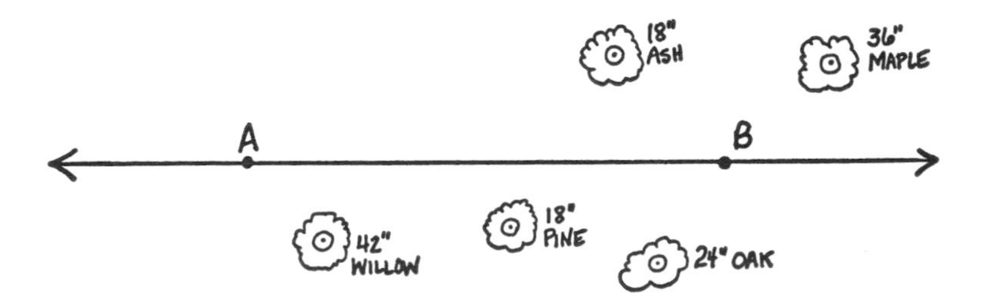

Using protractors and rulers, students record the required "field" note information. You may want to collect the drawings and require students to redraw their own maps from their field notes. The "Tree Map Example" handout that is provided with this project shows how to construct a tree map. A convenient scale, such as 1 inch = 40 feet, should be established.

b. Tree mapping can also be conducted on the floor of your classroom by clearing the desks to one side and running a string or cord the length of the room. This string becomes the base line, and points "A" and "B" are clearly marked. Be sure the string is tightly stretched and is firmly secured at both ends. "Trees" are placed around the room, represented by circles of green or brown construction paper taped to the floor. Students stretch string from a "tree" to one of the points on the base line and read the resulting angle from a properly positioned protractor. Distance is measured with tape measures, yardsticks, meter sticks, or whatever other method you choose.

c. The preceding instructions can be followed in a larger room, like a gymnasium. This will prove to be much less congested and students will have more freedom to work on their own.

- Originally this was designed as a tree identification project. The idea was for students to put together a tree map of a local park, showing where every *kind* of tree was located. Each time a tree was located geometrically it also had to be identified, and students were equipped not only with transits, measuring tapes, notepads, and pencils, but also with tree identification books and sample bags. You may want to emphasize tree identification also, or combine the project with a biology class in which students learn about tree identification.

Name ______________________________ Date ______________________________

PRACTICE FIELD NOTES

When a project (or an occupation) requires a person to go outside to collect information or do a job, the activity is called field work. The "field" is a place away from the classroom (or office) where the work is done. Archaeologists are doing field work when they go to a site to unearth an ancient artifact. Geologists are in the field when they study volcanoes. Many occupations require field work. For example, land surveyors, architects, oceanographers, private investigators, police, construction workers, anthropologists, lawyers, and engineers are all people who do field work as part of their jobs.

One of the most crucial parts of a field worker's job is taking notes so there is an accurate record of what is done or observed. These are called field notes. You will be taking field notes during this project, and it will become apparent as you work that the information gathered could *never* be memorized. It must be written down. A good set of field notes for this kind of activity includes a rough drawing with everything labeled, and a list of all the data that is collected. An example of what a set of field notes for the Tree Mapping project should look like will be provided by your teacher. Your data should be arranged in the same way.

FOR YOUR INFORMATION:

- The instrument used to measure angles outdoors is called a *transit.*
- In your notes, "transit point" means the point that the transit is set over. This point is the *vertex* of all angles measured from there.
- A base line is a line from which all angle measurements begin. For this project the two primary reference points, "A" and "B," are located on the base line. The angles to various trees can be measured by placing a transit over either of these two points.

Name ________________________ Date ________________________

FIELD NOTES EXAMPLE

Directions: When working in the field it is necessary to take accurate notes. For the tree mapping project you will record data that will allow you to draw a map showing the location of several trees in an area selected by your teacher. Your field notes should be recorded like those in the example below. Be careful to accurately record each piece of information as you measure angles and distances.

Distance $\overline{AB}$ = 500 ft
Distance $\overline{AC}$ = 273 ft
Distance $\overline{BC}$ = 227 ft

Important note: The "size" of a tree is its *diameter* in inches. This measurement should be estimated, not calculated.

Tree Letter	Size	Kind of Tree	Transit Point	Angle	Left or Right	Distance
D	24 in.	Red maple	A	BAD = 60°	Left	93 ft
E	18 in.	White pine	A	BAE = 37°	Left	137 ft
F	36 in.	Sycamore	A	BAF = 10°	Left	115 ft
G	12 in.	Sweetgum	A	BAG = 68°	Right	134 ft
H	24 in.	Red oak	C	BCH = 78°	Left	165 ft
I	72 in.	Weeping willow	C	ACI = 46°	Left	68 ft
J	8 in.	Redbud	B	ABJ = 76°	Right	168 ft
K	12 in.	Ginko	B	ABK = 32°	Right	106 ft
L	36 in.	Sugar maple	B	ABL = 39°	Left	103 ft
M	30 in.	Pin oak	B	ABM = 79°	Left	143 ft

Name ______________________ Date ______________________

TREE MAP EXAMPLE

Use this map as a guide for developing a tree map from your field notes.

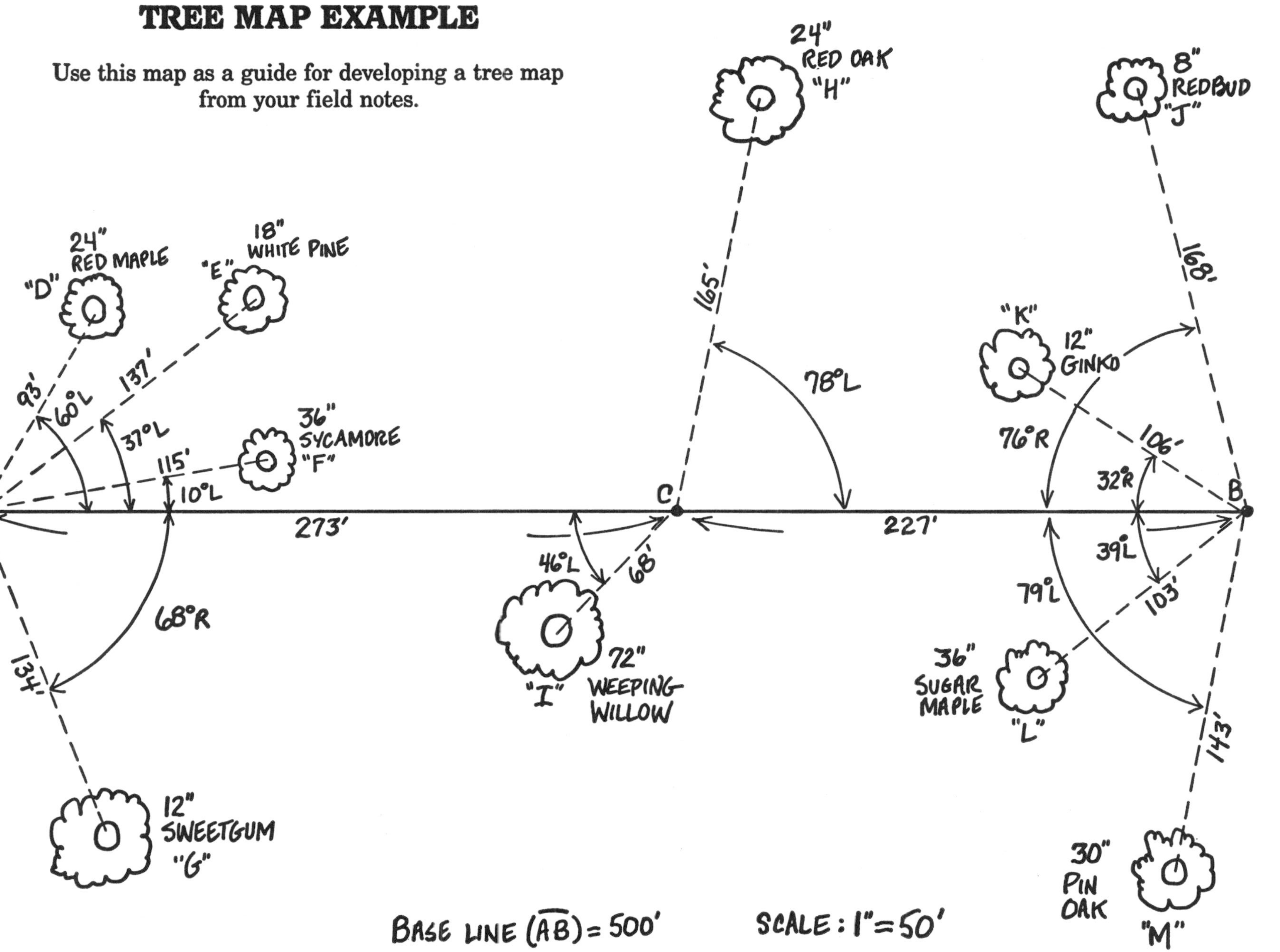

Name ______________________________ Date ______________________

TREE MAPPING
Student Assignment Sheet

There are many applications for geometry in the world. It is one of the most practical, useful branches of mathematics. The relationship among lines, angles, and distances is the basis for highway design, building construction, and even space flight. Mapping is also an undertaking that is dependent upon geometry, because an accurate map locates things in a given area or territory. Learning to make a tree map will give you insight into how geometry can be used. It will also introduce you to a procedure for collecting field notes. This kind of project can be fun to do, as well as educational. You will find that the skills needed for tree mapping will be useful in future mathematics and science classes; they may even help you in a future occupation.

This is a small group project (three to four people per group). Each person will turn in a complete set of field notes and a finished map, so be sure to do everything that is outlined in this handout. You will measure angles with a transit and measure distances either with a tape measure or by pacing. Here is the assignment:

I. Find the base line that has been marked outside. One point is marked "A" and the other point is marked "B."

II. Measure the distance between "A" and "B." Record this distance.

III. Select eight trees that can be seen from point "A" or point "B." (Later on, you will label each tree with a letter as you map it.) *Do not select trees that are more than 200 feet from the base line.*

IV. Select two additional trees that can be seen from a third point that you will locate somewhere on $\overline{AB}$ (between "A" and "B"). This will be point "C," and to locate it, follow these directions:

A. Set the transit over either point "A" or point "B," and sight the other point.

B. Put a mark on the base line (using your transit for line) at a place where both of the trees you have chosen can be seen.

C. This mark is point "C." Measure how far "C" is from "A" and "B." Record these distances. (They should add together to give the total distance from "A" to "B".)

V. Locate all ten trees by setting the transit over one of the three points on the base line, turning an angle to each tree, and then measuring its distance from the transit. Record the transit point (the point on the base line that the transit is over) for each tree you locate.

VI. Give each tree a letter, "D" through "M," to help with note taking and particularly with identifying angles. Record each tree's size (estimated diameter); you may also be asked to record the *type* of tree (maple, oak, pine, and so forth).

VII. Record whether each angle is turned to the left or to the right from the base line.

VIII. When all the necessary information is recorded, return to the classroom and begin work on a map. On the due date, hand in a copy of your group's field notes and a map that you have drawn. Your work must be neat and accurate to receive a good grade!

Name ______________________________ Date ______________________________

TRANSIT MATERIALS

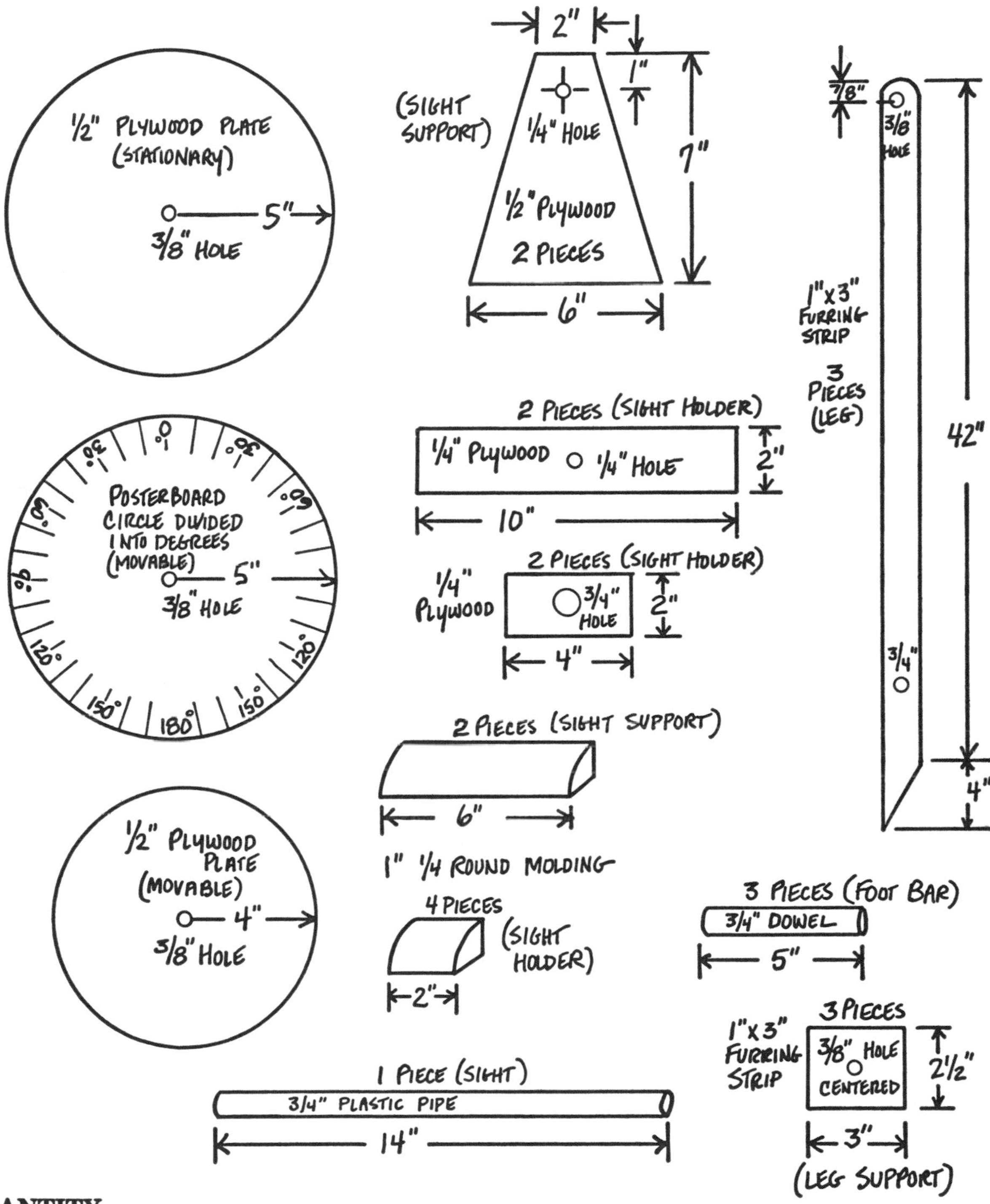

QUANTITY

(4) $\frac{3}{8}$ × 2 in. carriage bolts, washers, and wing nuts
(2) $\frac{1}{4}$ × 1 in. carriage bolts, washers, and wing nuts
(12) 8 × 1$\frac{1}{4}$ in. flathead wood screws, small brads, and glue

Name ______________________ Date ______________________

TRANSIT CONSTRUCTION DIAGRAM

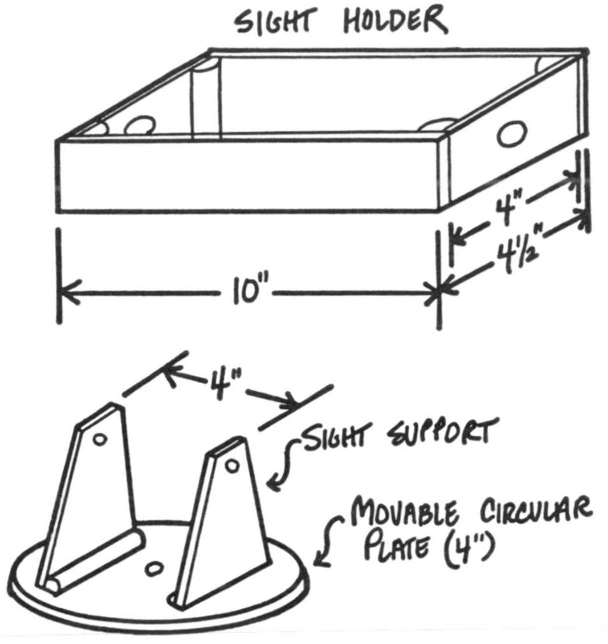

1. Assemble the $\frac{1}{4}$ in. plywood pieces for the sight holder with nails and glue. Corners are braced with $\frac{1}{4}$ round molding.
2. Attach the $\frac{1}{4}$ round molding to the $\frac{1}{2}$ in. plywood sight supports with nails and glue.
3. With screws and glue attach the supports to the 4 in. radius plywood plate (4 in. apart, outside to outside). Countersink screws. Outside of supports should be 2 in. from center of circle.

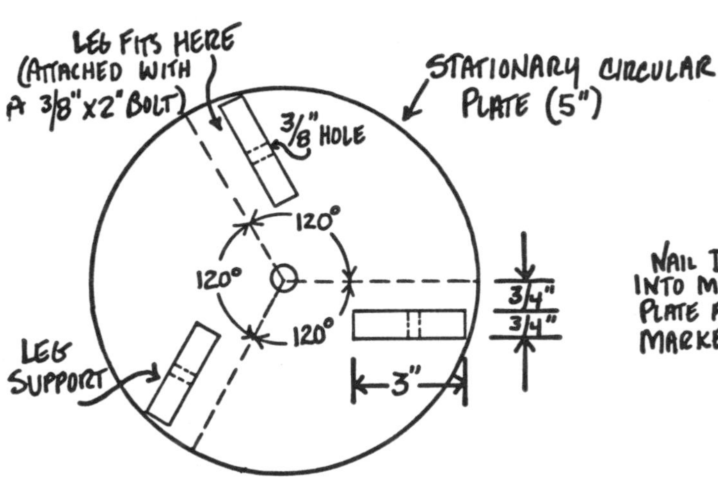

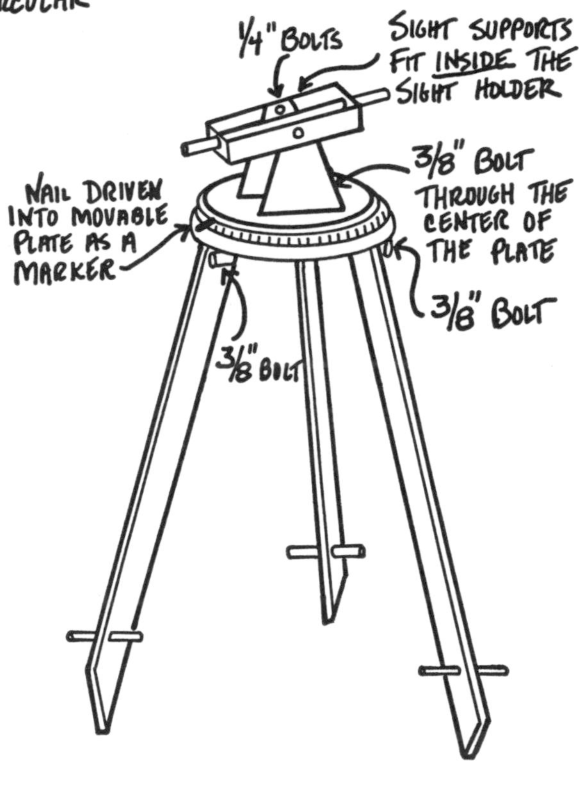

4. On one side of the 5 in. radius plywood plate, lay out the positions for the $2\frac{1}{2}$ in. × 3 in. pieces of furring strip. Make sure the pieces are glued and screwed on securely. The legs are then bolted on to these pieces.
5. The posterboard protractor is sandwiched between the plywood plates but must be free to turn.

M-7

LAND SURVEYING

Teacher Preview

General Explanation:

This project consists of a number of surveying problems that students either stake out to actual size in a playground, field, or park, or that they draw to scale on paper. It is described as an outdoor project.

Length of Project: 9 hours

Level of Independence: Intermediate

Goals:

1. To make use of point location skills learned in previous courses and projects.
2. To provide a mathematics project that emphasizes geometry, scale drawing, and following specific instructions.
3. To allow students to work independently to complete "land surveys."
4. To take students outside for an educational activity.
5. To provide an opportunity to teach students about land surveying.

During This Project Students Will:

1. Measure angles and distances from established points on a base line.
2. Follow specific survey instructions.
3. Produce accurate scale drawings of their surveys.
4. Conduct full-scale surveys with homemade transits (optional).

Skills:

Using a compass
Using a protractor
Using a ruler
Angular measurement
Collecting data
Neatness
Accuracy
Sense of "quality"
Self-confidence
Organizing
Using a homemade transit
Linear measurement
Persistence
Sharing space
Taking care of materials
Personal motivation
Drawing/sketching/graphing
Mapping
Scale drawing
Listening

Working in groups
Accepting responsibility
Concentration
Controlling behavior
Individualized study habits
Using a straightedge
Point location
Observing
Drawing straight and parallel lines
Following project outlines
Poster making

Handout Provided:

- "Student Assignment Sheet"

PROJECT CALENDAR:

HOUR 1: ______	HOUR 2: ______	HOUR 3: ______
An introduction to land surveying as an ancient and necessary occupation is given. Point location skills are reviewed. PREPARATION REQUIRED	The assignment sheet is handed out and students do the first two problems on drawing paper. NEED SPECIAL MATERIALS HANDOUT PROVIDED STUDENTS TURN IN WORK ANSWERS PROVIDED	The two problems done during Hour 2 are discussed. Students are divided into small groups and begin working on the final four problems on drawing paper. RETURN STUDENT WORK NEED SPECIAL MATERIALS ANSWERS PROVIDED
HOUR 4: ______	**HOUR 5:** ______	**HOUR 6:** ______
Small groups complete their drawings of problems C–F, and these are checked. The assignment for making outdoor surveys of these problems is given. NEED SPECIAL MATERIALS STUDENTS TURN IN WORK ANSWERS PROVIDED	Students do problem C outside. NEED SPECIAL MATERIALS	Students do problem D outside. NEED SPECIAL MATERIALS
HOUR 7: ______	**HOUR 8:** ______	**HOUR 9:** ______
Students do problem E outside. NEED SPECIAL MATERIALS	Students do problem F outside. NEED SPECIAL MATERIALS	The project and the skills that it emphasized are discussed.

Lesson Plans and Notes

HOUR 1: Introduce students to land surveying as an important profession based upon point location. Spend the hour reviewing point location skills: how to set a base line and measure angles and distances to locate other points. Present several simple problems on the board to illustrate these skills. See the project in this book titled Protractor–Compass–Ruler.

HOUR 2: Give students the first two of six problems from the assignment sheet to do on drawing paper in the classroom. These are handed in at the end of the hour. Students need protractors, rulers, and drawing paper for Hours 2, 3, and 4.

Notes:

- Land surveying is an excellent desk project, so if you have no intention of organizing an outdoor activity, students can still do the surveys on paper. Each survey begins from a scale drawing of a 100-foot base line that is "above" the area to be surveyed. In other words, the base line is at the *top* of the paper on which the survey is being drawn. Point A is always on the *left* side of this line and point B is always on the *right* side. This arrangement is crucial for such problems as D, E, and F from the land surveying handout, where a point is located from the base line and then used as the beginning point for a survey.
- Teacher answer sheets that show what each survey should look like are provided after the student assignment sheet.

HOUR 3: Return the two surveying problems and discuss common mistakes. Divide students into small groups (two to four students per group), and assign them the remaining four problems to study and make drawings of in class.

HOUR 4: Small groups complete their drawings, which are quickly checked to ensure that each group knows what the surveys will look like. Tell students that they will do one survey each hour for the next four hours. If transits have been built, students will do these surveys outside. Each group will need a transit, a clipboard, measuring tape, chaining pins, and stakes with ribbons (one per corner of the survey). If you do not have transits, the surveys can be done on a large floor space or on posterboard. Appropriate scales must be established for indoor surveys.

HOUR 5: Students do problem C outside.

HOUR 6: Students do problem D outside.

HOUR 7: Students do problem E outside.

HOUR 8: Students do problem F outside.

HOUR 9: After the surveys are complete, set aside an hour for discussion of the project and the skills that were employed. Give special attention to particular difficulties that were encountered during the outdoor portion of the project. Ask students if they can invent, or create, surveying problems of their own. This shows that once skills are learned, they can be applied to specific situations. Also discuss the level of independence needed for the project: what is required of students who work on projects like this on their own? Emphasize the need for independent learning skills in any situation where students are allowed to learn on their own.

General Notes About This Project:

These notes pertain to an outdoor, full-scale land surveying project: first, transits must be built. See the Tree Mapping project in this book for plans, notes, and suggestions about building and using a homemade transit.

- The assignment sheet included with this project supplies six land surveying problems for students to solve. The first two are done on paper in class for practice. The remaining four problems are done outside, one per hour, but only after students draw them on paper in class so they know what each looks like. You may decide to change these requirements and have students do only 1, 2, or 3 problems outside in actual surveys if time is limited.
- This project works best if each group has its own base line from which to work.
- All of the angles should be turned to the same side of the base line ($\overline{AB}$). In other words, one angle will not be above the base line if the others are below the line. In the field, $\overline{AB}$ should be placed so that all available space is on one side of it; on paper, the base line should be drawn as shown on the student handout.
- There are a number of instructions that read like this: "angle BFG = 42° left." The first letter is a point that has already been set and is being sighted. The middle letter is always the vertex, where the transit sits. The third letter is the point that the angle is being turned to, or the point that is going to be set. This means that points "B" and "F" are in the ground (or on paper), and that the transit is over point "F" (the vertex of the angle) and is sighting point "B." An angle of 42° is to be turned *to the left* to locate point "G." This is very important and must be thoroughly covered in class before actual surveying begins. If angle BFG simply equals 42°, it can be turned right or left and be equally correct. If these instructions for turning angles are not carefully followed, the surveys will go awry.
- You will need a quantity of wooden stakes and hammers for marking survey corners. Colored plastic ribbon will also help make the stakes visible. When the survey is done on pavement use chalk, crayon, or highway cones, if they are available.
- Something should be provided to mark, or "hold," line and distance in the field. Surveyors call these "chaining pins," but 16 penny nails with colored plastic ribbon tied to their heads work very well, as do quarter-inch dowel rods cut 12 inches long and sharpened on one end.

- When intersecting two angles, place chaining pins at short intervals on the line of one angle (while sighting with a transit), and then intersect this line with the second angle by resetting the transit and sighting between two of the pins. A string can be stretched between these two pins for exact line in one direction while the transit provides line from the other direction. On pavement do this with chalk or crayon.
- You may want to provide kite string for students to stretch between corners so it is easier to visualize the area they have surveyed.
- This project deals with large areas and the possibility for error is great. All angles have been rounded to the nearest degree and all distances have been rounded to the nearest foot. Better accuracy cannot be obtained with the equipment being used. In the field the surveys may have only a remote resemblance to the drawings that are provided. Precision and careful accuracy are very important, but even with great care the results will not be perfect. Evaluate students on *procedure.* If they do everything as they are supposed to, their results will be fairly close, but even a small mistake can sometimes throw things off by quite a bit.
- So that survey points will be clearly identified, be sure to have each group mark its stakes with the proper letters before putting them in the ground. This must be done to avoid confusion, and so you can quickly determine if they have located their points properly. Colored notecards with large letters thumbtacked to wooden stakes works well for this.
- An alternative way of having students do outdoor surveys is to give them a drawing of the survey instead of written instructions. They first measure angles with protractors and distances with rulers, use a scale to change distances to real dimensions and record their measurements in field-note form; then they go outside to conduct the survey.
- As with other outdoor activities, this project runs more smoothly if you have additional adult help.

Name ________________________ Date ________________________

LAND SURVEYING
Student Assignment Sheet

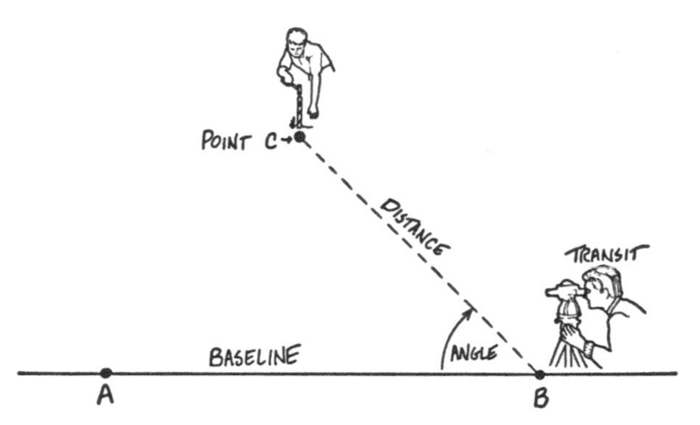

Land surveying is an old American institution—and a historically vital occupation. George Washington was a land surveyor. After the Louisiana Purchase, Thomas Jefferson decreed that the land would be divided into parcels of land one mile on a side, and the surveyors moved in to do the job. When the railroads were built, surveyors led the way; the same is true of the interstate highways. Today, hardly a house is built or a building constructed without first having the piece of ground it will sit on surveyed. It makes sense that in a country where property is privately owned, an essential job would be establishing accurate property corners, rights of way, boundary lines, and permanent markers; these are some of the things a surveyor does.

For this project you will work either with a partner or in a small group to conduct surveys. This handout gives several sets of instructions for imaginary land surveys. You may be asked to go outside and stake these surveys in an open area, make scale drawings of them on paper, or do both. If you wish, include extra details in your drawings, such as houses, streams, islands, and so forth. Here are your instructions:

I. When making a drawing, establish a base line at the "top" of your survey with point "A" on the left side of the line and point "B" on the right. This line represents a 100-foot base line. Use a scale of 1 inch = 10 feet on your survey drawings. This means that the base line will be 10 inches long, and will be arranged on the drawing paper as shown below.

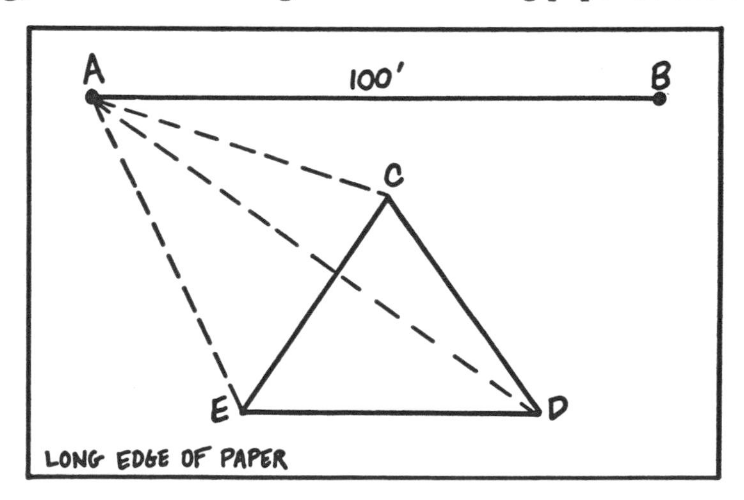

Note: A scale of 1 inch = 10 feet means that $\frac{1}{2}$ inch = 5 feet, $\frac{1}{4}$ inch = $2\frac{1}{2}$ feet and $\frac{1}{8}$ inch = $1\frac{1}{4}$ feet. You will have to estimate to the nearest foot when you make your drawings. For example, a measurement of 54 feet would be scaled down to 5 inches (50 feet) plus $\frac{3}{8}$ inches (approximately 4 feet).

Name ______________________________ Date ____________________

LAND SURVEYING

Student Assignment Sheet (continued)

II. When doing a full-scale survey outside, follow the teacher's instructions carefully. This will be a test of your ability to operate on your own.

III. Problems

A. Points "A" and "B" are the front corners of a city lot that has a house on it. Find the corners of the house by setting a stake at each of the points described below.

1. Corner "C": angle ABC = 8°; $\overline{BC}$ = 85 ft
2. Corner "D": angle ABD = 10°; $\overline{BD}$ = 70 ft
3. Corner "E": angle ABE = 22°; $\overline{BE}$ = 75 ft
4. Corner "F": angle ABF = 31°; $\overline{BF}$ = 54 ft
5. Corner "G": angle ABG = 33°; $\overline{BG}$ = 99 ft
6. Corner "H": angle ABH = 48°; $\overline{BH}$ = 70 ft

Connect points: C and D, F and H
D and E, G and H
E and F, G and C

B. Points "A" and "B" are the front corners of a lot where a grocery store is going to be built. Find where the corners of the store will be by setting a stake at each of the points described below.

1. Corner "C": angle BAC = 5°; $\overline{AC}$ = 34 ft
2. Corner "D": angle BAD = 39°; $\overline{AD}$ = 44 ft
3. Corner "E": angle BAE = 62°; $\overline{AE}$ = 32 ft
4. Corner "F": angle BAF = 76°; $\overline{AF}$ = 60 ft
5. Corner "G": angle ABG = 5°; $\overline{BG}$ = 34 ft
6. Corner "H": angle ABH = 39°; $\overline{BH}$ = 44 ft
7. Corner "I": angle ABI = 62°; $\overline{BI}$ = 32 ft
8. Corner "J": angle ABJ = 76°; $\overline{BJ}$ = 60 ft

Connect points: C and G, I and J, E and D
G and H, J and F, D and C
H and I, F and E

C. Points "A" and "B" are on a base line that is on one side of a 15-foot stream; the four corner poles for a "pole" barn have been set on the other side of the stream. You have to locate each corner by intersecting angles from point "A" and point "B." Set a stake at each of the points described below.

1. Corner "C": angle BAC = 13°; angle ABC = 59°
2. Corner "D": angle BAD = 49°; angle ABD = 14°
3. Corner "E": angle BAE = 74°; angle ABE = 36°
4. Corner "F": angle BAF = 35°; angle ABF = 78°

Connect points: C and F, E and D
E and F, D and C

D. Point "A" and point "B" are on a small island. You will first find point "C" by intersecting two angles. The survey is then started from point "C," which is on the shore of another, smaller island. From point "C," find a diamond shaped area that has a treasure buried in it. Set a stake at each point of the diamond. If you connect the opposite points of the diamond (so that the lines intersect in the interior of the diamond) you will have found Captain Red Beard's treasure, and you will have proven your ability as a land surveyor.

1. Point "C": angle BAC = 20°; angle ABE = 20°;

Name ______________________________ Date ______________________

LAND SURVEYING
Student Assignment Sheet (continued)

2. Point "D": angle BCD = 40° Right; $\overline{CD}$ = 17 ft
3. Point "E": angle ACE = 61° Left; $\overline{CE}$ = 39 ft
4. Point "F": angle BCF = 92° Right; $\overline{CF}$ = 48 ft
5. Point "G": angle BCG = 56° Right; $\overline{CG}$ = 43 ft

Connect points: D and G, F and E
G and F, E and D

E. Points "A" and "B" are on the shore of an island that is being developed for vacationers. First locate point "C," then find the corners of the luxury vacation cottage that is planned for that area. Set a stake at each of the points described below:

1. Point "C": angle BAC = 39° angle; $\overline{ABC}$ = 39°
2. Point "D": angle BCD = 25° Left; $\overline{CD}$ = 34 ft
3. Point "E": angle BCE = 8° Left; $\overline{CE}$ = 23 ft
4. Point "F": angle BCF = 15° Right; $\overline{CF}$ = 38 ft
5. Point "G": angle BCG = 46° Right; $\overline{CG}$ = 35 ft
6. Point "H": angle BCH = 55° Right; $\overline{CH}$ = 16 ft
7. Point "I": angle BCI = 91° Right; $\overline{CI}$ = 24 ft
8. Point "J": angle ACJ = 91° Left; $\overline{CJ}$ = 24 ft
9. Point "K": angle ACK = 55° Left; $\overline{CK}$ = 16 ft
10. Point "L": angle ACL = 46° Left; $\overline{CL}$ = 35 ft
11. Point "M": angle ACM = 15° Left; $\overline{CM}$ = 38 ft
12. Point "N": angle ACN = 8° Right; $\overline{CN}$ = 23 ft
13. Point "O": angle ACO = 25° Right; $\overline{CO}$ = 34 ft

Connect points: O and D, G and H, K and L
D and E, H and I, L and M
E and F, I and J, M and N
F and G, J and K, N and O

F. Points "A" and "B" are on a base line that has been set to help survey the perimeter of a new office building. Two of the building's corners will be located by turning an angle from point "A" and measuring the proper distances. The rest will be located by putting the transit over a corner, sighting the last corner set, turning an angle to the next corner, measuring the proper distance, setting the corner, and then repeating the procedure. Set a stake at each of the points described below:

1. Point "C": angle BAC = 45°; $\overline{AC}$ = 14 ft
2. Point "F": angle BAF = 45°; $\overline{AF}$ = 85 ft

Now the starting point becomes point "F." Follow the rest of the instructions very carefully:

3. Point "G": angle AFG = 102° Right; $\overline{FG}$ = 36 ft
4. Point "H": angle FGH = 124° Right; $\overline{GH}$ = 30 ft
5. Point "I": angle GHI = 90° Right; $\overline{HI}$ = 20 ft
6. Point "J": angle HIJ = 117° Right; $\overline{IJ}$ = 23 ft
7. Point "K": angle IJK = 117° Left; $\overline{JK}$ = 20 ft
8. Point "L": angle JKL = 117° Left; $\overline{KL}$ = 23 ft
9. Point "E": angle AFE = 45° Left; $\overline{FE}$ = 20 ft
10. Point "D": angle FED = 145° Left; $\overline{ED}$ = 36 ft

Connect points: I and H, E and D K and J
H and G, D and C, J and I
G and F, C and L,
F and E, L and K

TEACHER ANSWER SHEET:
Land Surveying Problem A

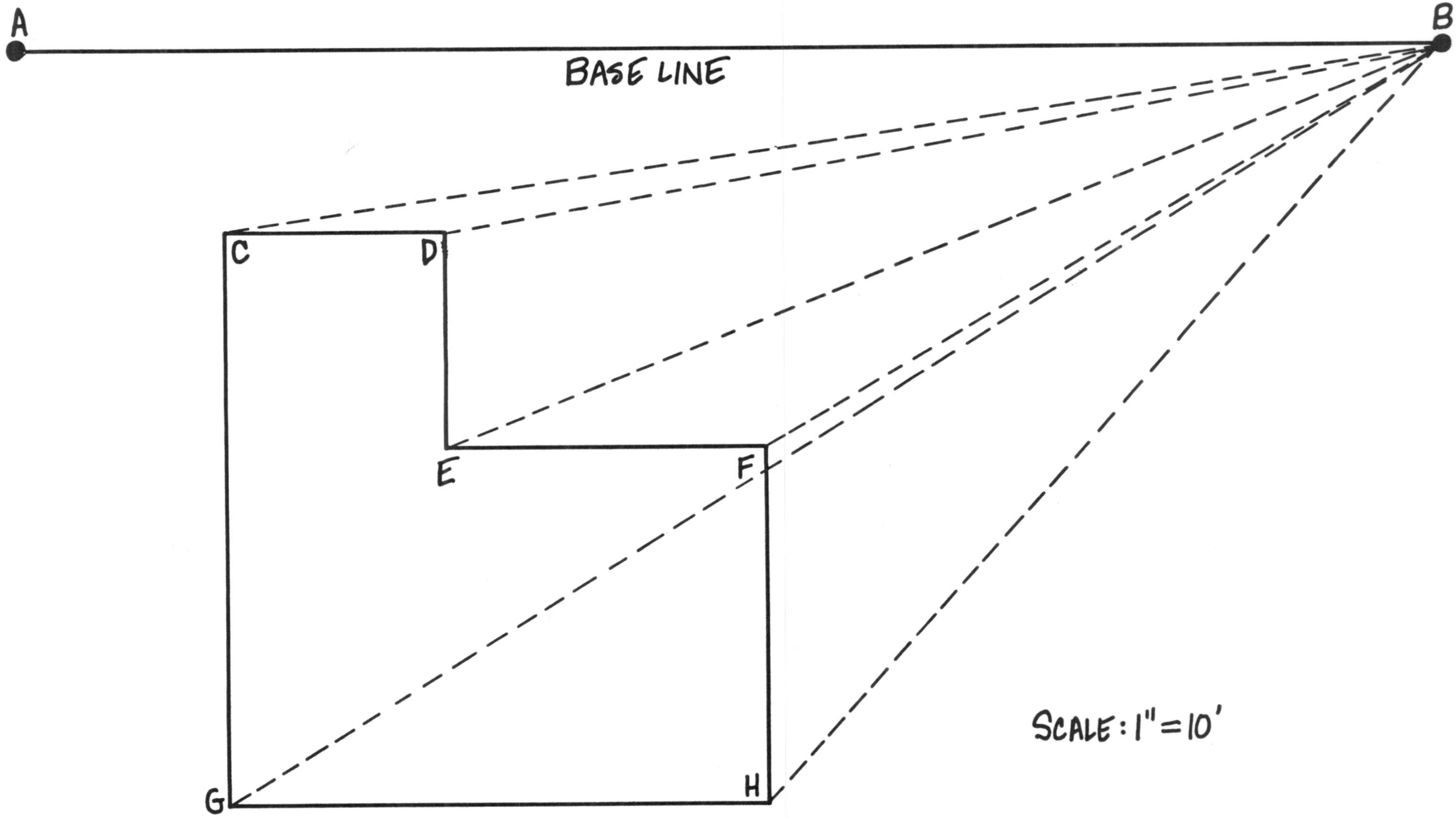

A

B

BASE LINE

C

D

E

F

G

H

SCALE: 1" = 10'

TEACHER ANSWER SHEET:
Land Surveying Problem B

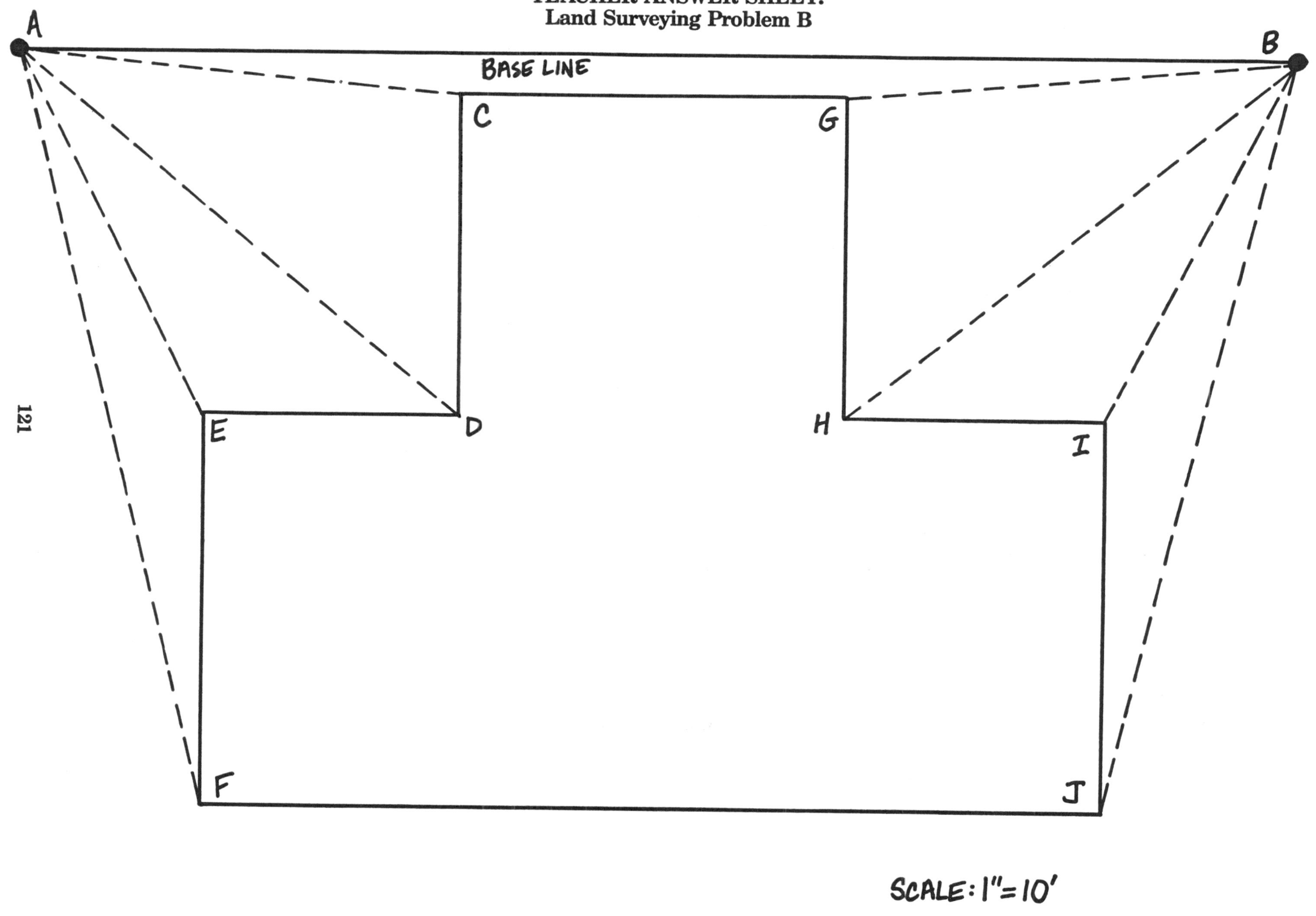

TEACHER ANSWER SHEET:
Land Surveying Problem C

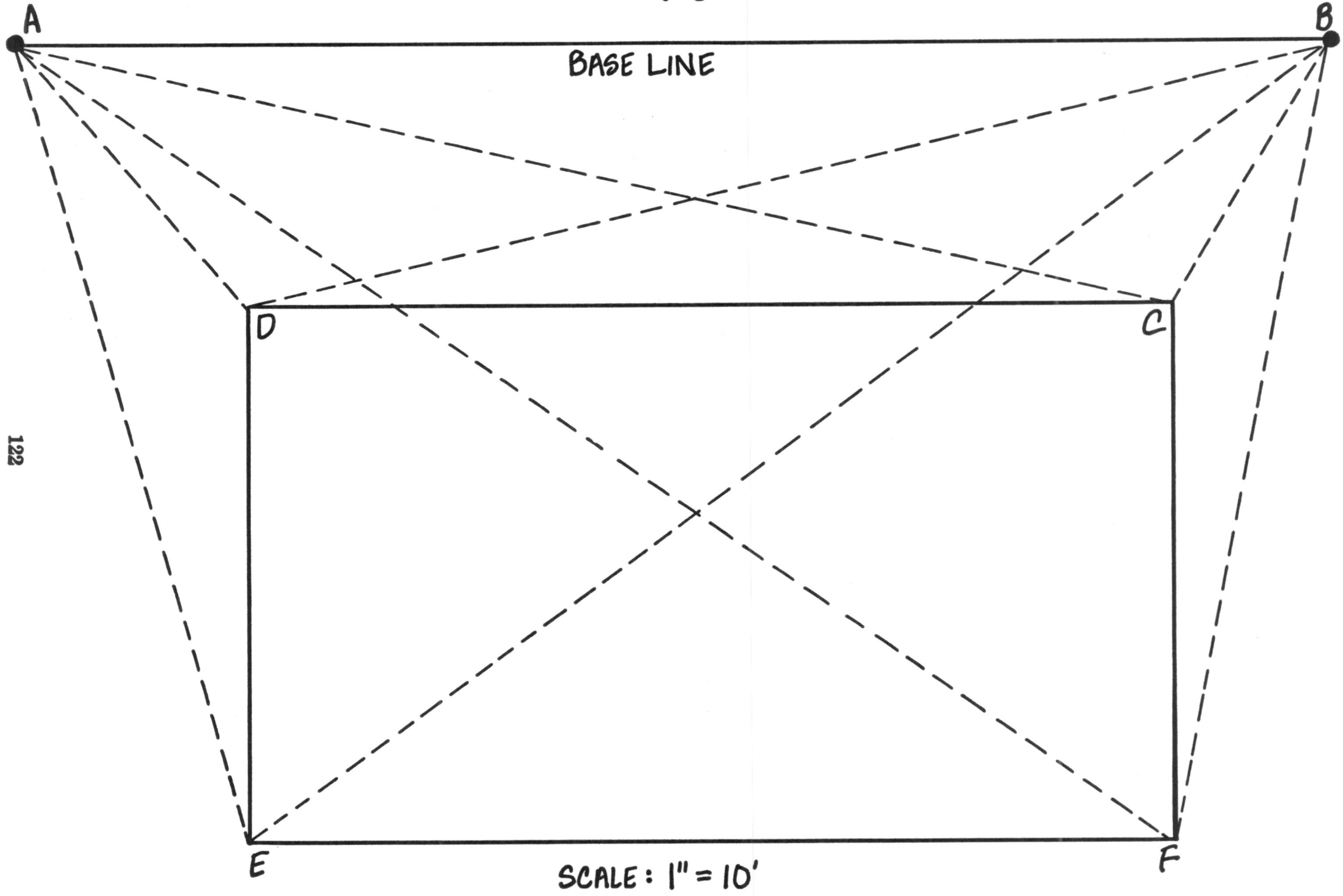

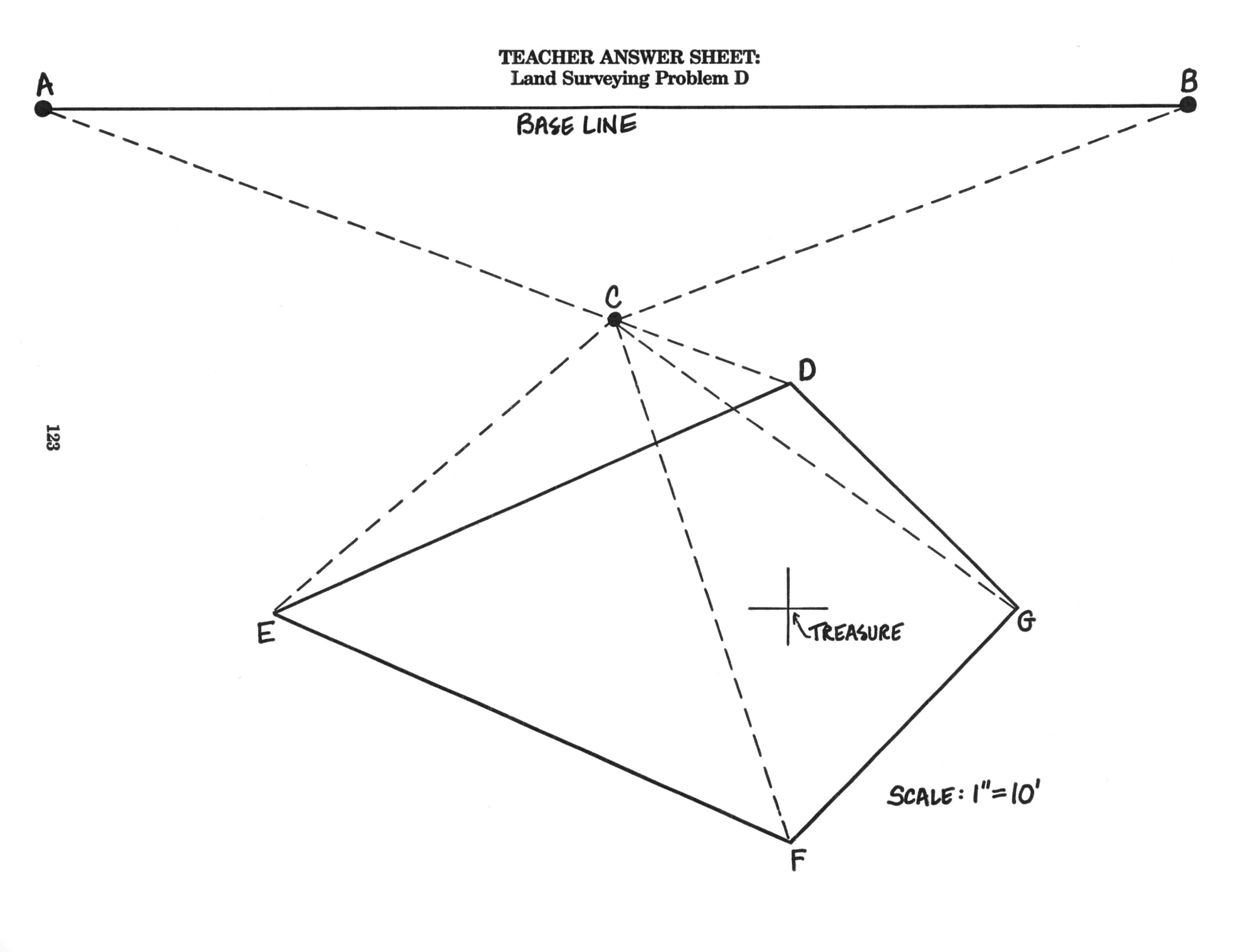
TEACHER ANSWER SHEET:
Land Surveying Problem D
A
B
BASE LINE
C
D
E
F
G
TREASURE
SCALE: 1"=10'

TEACHER ANSWER SHEET:
Land Surveying Problem E

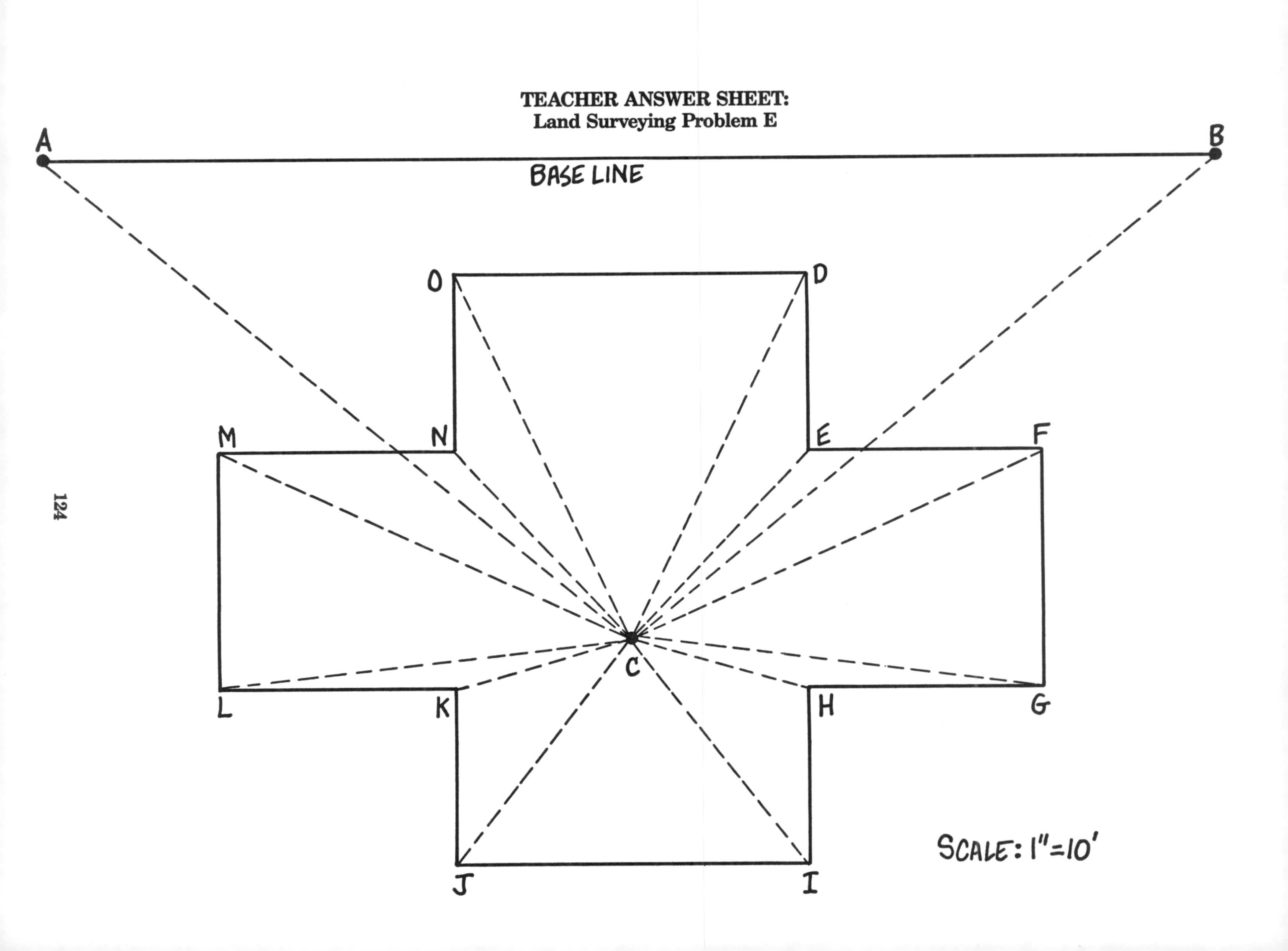

TEACHER ANSWER SHEET:
Land Surveying Problem F

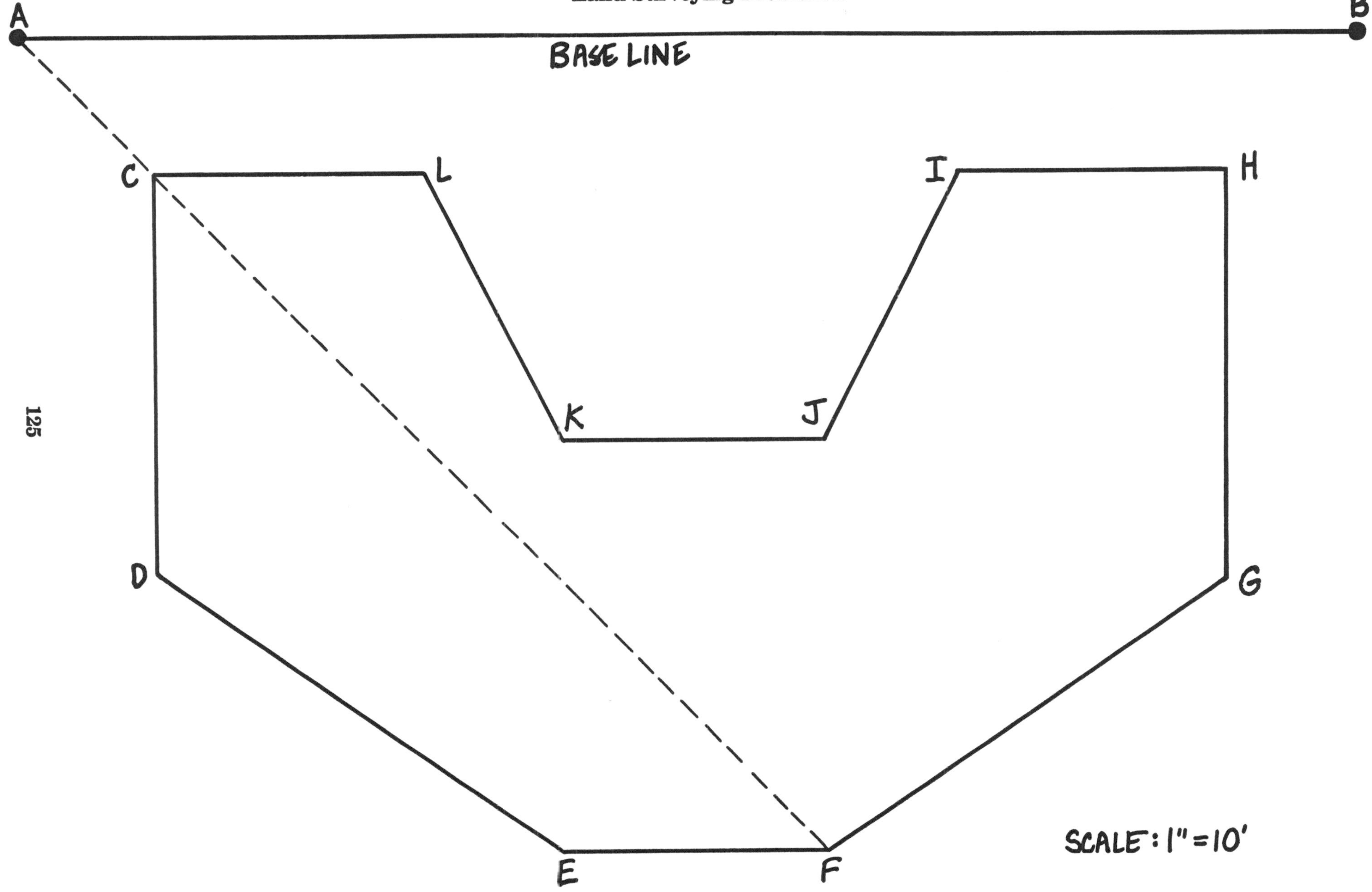

A B
BASE LINE
C L I H
K J
D G
E F
SCALE: 1" = 10'

M-8

MATHEMATICS IN THE PARK (TREE MEASUREMENT)

Teacher Preview

General Explanation:

This project is concerned primarily with tree measurement: height, circumference, and volume. It is specifically designed to provide an outdoor mathematics activity for students. The project can be conducted with just a few trees; if a park is not accessible, a neighborhood or school yard that has as few as four or five trees will suffice. Students use equations to make calculations after going outside to collect data. Homemade clinometers are used to estimate the height of trees. Instructions are provided for making simple clinometers.

Length of Project: 7 hours

Level of Independence: Intermediate

Goals:

1. To combine the study of mathematics with an outdoor activity.
2. To place emphasis on studies of nature and natural things.
3. To introduce students to the use of field data and standard equations to discover or calculate numbers that have real meaning.
4. To introduce students to the concept of using triangles to calculate unknown distances.

During This Project Students Will:

1. Demonstrate correct use of basic arithmetic skills and simple equations.
2. Calculate their average pace lengths.
3. Build and demonstrate proper use of clinometers.
4. Calculate the area of circles.
5. Calculate the volume of cylinders.
6. Calculate the number of board feet in tree trunks and their volume in cubic feet.

Skills:

Collecting data
Listening
Neatness
Personal motivation
Sense of "quality"
Basic mathematics skills

Organizing
Working with limited resources
Accepting responsibility
Controlling behavior
Individualized study habits
Taking care of materials
Time management
Following project outlines
Using equations
Solving for an unknown
Observing
Accuracy
Identifying problems
Concentration
Persistence

Handouts Provided:

- "Wood for Fuel Worksheet"
- "Pacing Worksheet"
- "Tree Measurement Student Assignment Sheet"
- "Tree Measurement Student Answer Sheet"
- "Mathematics in the Park Final Test"
- "Mathematics in the Park Final Test–Student Answer Sheet"
- "How to Build a Clinometer"
- "Clinometer Scale"

PROJECT CALENDAR:

HOUR 1: ________	HOUR 2: ________	HOUR 3: ________
Students do the two problems on the "Wood for Fuel Worksheet." During the last part of the hour they calculate their average pace length. HANDOUTS PROVIDED	Clinometers are built in class. A clinometer scale and instructions for building a clinometer are provided. HANDOUTS PROVIDED PREPARATION REQUIRED NEED SPECIAL MATERIALS	Clinometers are explained and assignment sheets are distributed. Sample problems are worked on the board to show how data is recorded and how calculations are done. PREPARATION REQUIRED HANDOUTS PROVIDED
HOUR 4: ________	**HOUR 5:** ________	**HOUR 6:** ________
Students go outside to measure trees and record data on their assignment sheets. A due date is set for answer sheets to be turned in.	Graded answer sheets are returned and common mistakes are discussed. Samples of student work are done on the board as a review for the final test. RETURN STUDENT WORK	Students take the final test. HANDOUT PROVIDED
HOUR 7: ________	**HOUR 8:** ________	**HOUR 9:** ________
Final tests are returned and discussed. The discussion then turns to the usefulness and applications of the mathematics involved in this project. RETURN STUDENT WORK		

Lesson Plans and Notes

HOUR 1: Spend the first part of the hour discussing and working the two problems on the "Wood for Fuel Worksheet." Have students spend the rest of the hour calculating the average length of their paces. This information will be applied later when they use clinometers to help calculate how much wood is in certain trees. A worksheet for the pacing activity is provided.

Notes:

- A clinometer is an instrument for measuring the height of an object when one stands a certain distance from it. The height is read directly from a scale.
- The clinometer scale provided with this project is based upon a reading taken 100 feet from the base of the tree. It is important to accurately determine the length of each student's pace so that this distance can be measured without a tape.
- The answers to most of the problems in this project will be different for each student, since data is independently collected; therefore a teacher answer sheet cannot be provided. However, answers to the two problems on the "Wood for Fuel Worksheet" will be the same for everyone. They are as follows:

WOOD FOR FUEL WORKSHEET ANSWERS

Answer 1: 384 ft^3 $(l \times w \times h) \times 3$

Answer 2: 254.3 ft^3 $(\pi \times r^2 \times h)$

(approximately 2 cords of wood: Number of cords = $V \div 128ft^3/cord$)

HOUR 2: Students build clinometers that will be used to determine how much *usable* wood is in some of the trees in a local park (or around school or near students' homes). (The scale for the clinometers and instructions for building them are provided at the end of this project.)

HOUR 3: Teach students how to use a clinometer to measure the height of a tree. Then hand out the assignment sheet and the "Tree Measurement Student Answer Sheet" and work a few teacher-made sample problems on the board. Come to class prepared with three or four problems to use as examples. Here are some points to be made:

1. The clinometer is designed to be used 100 feet from the base of the tree being measured. This distance is *paced* by students (using the average length of pace calculated during Hour 1).
2. The useful wood in a tree ends where the trunk stops being straight or where it branches into different directions (or when it becomes too small). Some judgment is required to decide how high to sight on each tree.

3. The height is read directly off the clinometer scale when the observer is 100 feet from the tree. For other distances from the tree the height must be calculated (an equation is provided on the scale). Be sure to have students add the distance from the ground to eye level to their measurement of the height of each tree.
4. In order to calculate the volume of wood in a tree, its circumference must also be known. This is obtained by stretching a string around the trunk and then measuring the string. An eight-foot string will measure trees up to 30 inches in diameter.
5. These are the steps needed to calculate the volume of a tree trunk after finding its height and circumference:

a. Diameter: $c = \pi \times d$ (circumference = pi times diameter)
$d = \frac{c}{\pi}$ (this must be converted to *feet*)

b. Radius: $d = 2 \times r$
$r = \frac{d}{2}$ (in *feet*)

c. Area: $A = \pi \times r^2$ (in ft^2)

d. Volume: $V = A \times h$ (in ft^3)

6. There are 12 board feet in one cubic foot of wood, so
board feet = $V \times 12$ bd ft/ft^3

Tell the students that during the next hour they will be outside measuring trees. If it is impossible for the class to go outside, this should be done as a homework assignment. They should be careful and accurate in their data collection. Show the students how to record field notes in the appropriate spaces on their assignment sheets. Each person's calculations should be done on notebook paper and the answers recorded on the "Tree Measurement: Student Answer Sheet." The calculations should then be attached to the answer sheet before it is handed in.

Notes:

- When using a clinometer, a student is actually measuring the height of the tree *from eye level.* To get the real height, the distance from the ground to the eye must be added to the clinometer reading. This distance should be rounded to the nearest foot: 4 feet, 5 feet, or 6 feet.
- The point on a tree where "usable wood" ends and useless wood begins is a matter of judgment. This project works best with trees that are at least 18 inches in diameter and that have at least 8 feet of usable wood. Tree trunks should be fairly straight, without any splits, "Y" branches, or double trunks. Tell students to estimate where the trunk becomes definitely smaller than 18 inches thick, and sight that spot. When a trunk splits into two or more branches, the end of usable wood is obvious.

HOUR 4: Students go outside to measure trees. After they have taken and recorded the measurements, students come back to the classroom to begin working on the calculations. They will not be able to complete their calculations during this hour, so the work that remains to be done at the end of the hour becomes a homework assignment, due at a time specified by you. After collecting the "Tree Measurement: Student Answer Sheets" on the due date, give yourself time for grading before conducting the fifth hour of instruction.

Note:

- In some cases students may need more than one hour to measure ten trees. Be prepared to add an additional hour to the project schedule if necessary.

HOUR 5: Return the graded answer sheets and calculations to the students and explain common mistakes. Put some examples of student work on the board and critically examine them.

HOUR 6: Final test. The test calls for students to go outside and measure one tree, then return to the room to finish the test. It may be more convenient to give students these vital statistics for a tree in class and not ask them to go outside. This, however, does not test the use of a clinometer.

HOUR 7: Return and discuss the graded final tests. Spend the remainder of the hour discussing the mathematics involved in this project. It incorporates algebra and equations; it places emphasis on geometry, especially on calculating the area of a circle and the volume of a cylinder; it serves as an introduction to similar triangles, and even trigonometry, if an explanation of how the clinometer works is included in the project. In addition, the use of field notes to collect data is an important part of the project that has applications in many occupations and areas of study.

General Notes About This Project:

- This project can be expanded to have students do more work outside, and it can be combined with Sections and Acres or Tree Mapping to have students do field work to estimate how much wood is in a given area. Other activities, such as working with similar triangles, tree identification, discussion of tree care and uses of wood, and studies of the hardness of various types of wood, can be incorporated to give this project more scope.
- There can be no Teacher Answer Sheet for this project, since everyone's measurements will be different. This puts an added burden on you to calculate each answer. This is where a computer (or even a good calculator) comes in handy. A simple program can instantly give d, r, r^2, A, V, and *bd ft* when you enter h and c for each problem.

Name ______________________________ Date ______________________________

WOOD FOR FUEL WORKSHEET

Wood has been used as a fuel for fires since the beginning of humanity. Many people still heat their homes and cook with wood fires in furnaces, stoves, and fireplaces. The standard measure for wood that has been cut and stacked is the "cord." An underground house needs only about one cord of wood to keep it warm through one winter in a northern state like Michigan. In the same state an uninsulated farmhouse might require up to ten cords of wood to keep it warm. One of the reasons for measuring trees is to estimate how much wood they can provide for fuel.

Study the equations and then answer the two questions below. Remember that this project emphasizes calculating the area of circles and the volume of cylinders. Later you will learn how to find the radius of a tree trunk and its height.

Equations

area = length × width (rectangle)
area = $\pi \times r^2$ (circle)
volume = area × height (rectangular solid and cylinder)
volume = (length × width) × height (rectangular solid)
volume = $(\pi \times r^2) \times$ height (cylinder)

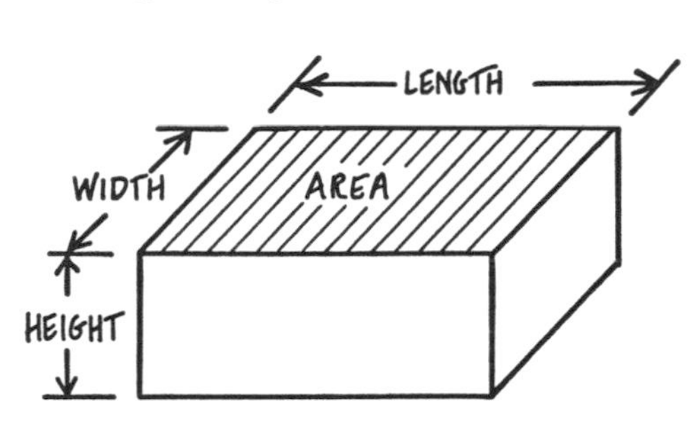

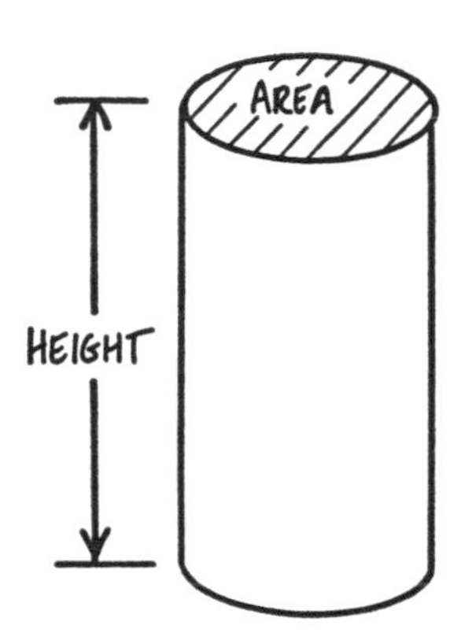

1. One cord of wood is 8 feet long, 4 feet wide and 4 feet high. How many cubic feet of wood are in 3 cords?

2. The usable height of a particular maple tree is 36 feet. The radius is 1.5 feet. How many cubic feet of usable wood are there in this tree? How many cords of wood are in the trunk of this tree? (approximately)

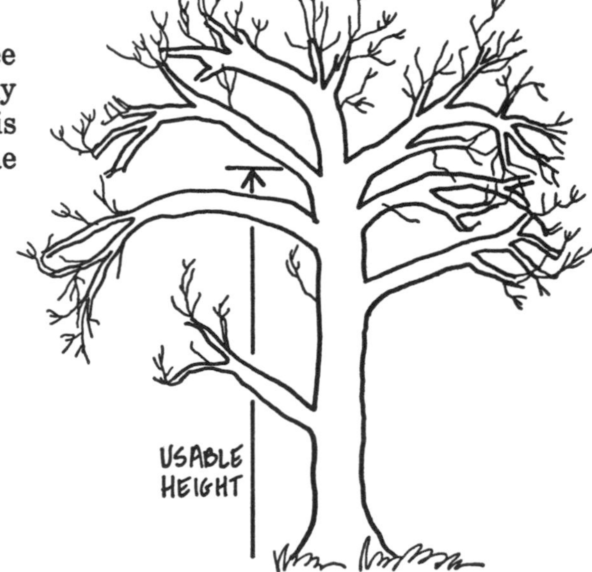

Name ______________________ Date ______________________

PACING WORKSHEET

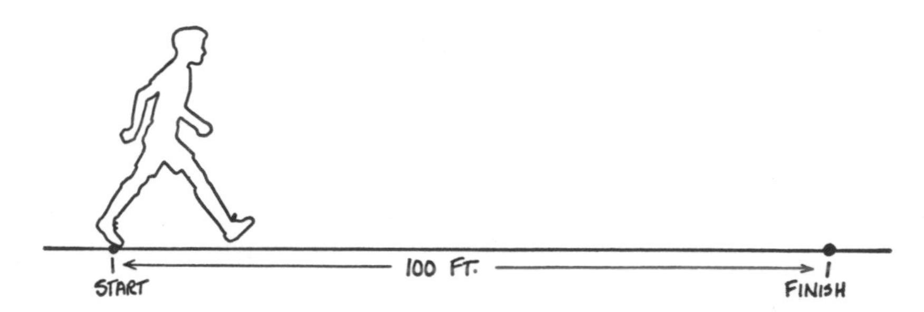

Pacing is one of the most basic forms of measurement. If you practice walking in a natural manner, taking the same length step each time, and if you know the average length of your stride, pacing can be a fairly accurate method of measurement. For the Tree Measurement project, an instrument called a clinometer is used to estimate the height of trees. The scale on the clinometer is accurate only if it is used at a distance of 100 feet from the base of the tree being measured. The purpose of this activity (pacing) is to ensure that you can pace 100 feet, and to establish the average length of your pace.

Follow the steps below and record the results in the spaces provided.

1. With a tape measure, measure 100 feet on pavement or flat ground.
2. Walk in a natural manner from one end of the 100 feet to the other—count each step as *one* pace.
3. Repeat three times and find the average. Record your data below.
4. Use this equation to find your length of pace:

$$\text{length of pace} = \frac{\text{distance walked (in this case, 100 feet)}}{\text{average number of paces}}$$

Number of paces in 100 feet

1	
2	
3	
Total	
Average number of paces in 100 feet	
Length of pace	

Do your calculations below. Be neat!

Name ______________________ Date ______________________

MATHEMATICS IN THE PARK (TREE MEASUREMENT)
Student Assignment Sheet

This project emphasizes geometry, primarily calculations for finding the area of a circle and the volume of a cylinder. It also makes use of the relationships between similar right triangles (that's how the clinometer works), and demonstrates how to calculate the number of board feet in the trunk of a tree. Mathematics in the Park offers an opportunity to go outside to collect data. With that opportunity comes the responsibility to follow instructions and to record information carefully. Most of your calculations will be done in the classroom or at home, so it is very important to have reliable data from which to work. Follow the directions below.

Select ten trees that contain at least one sawlog (18 inches in *diameter* and 8 feet long). Measure the circumference and height of each, and record the field data below. Put your calculations on notebook paper and hand them in with your answer sheet. A ten-foot piece of string, a yardstick, and a clinometer are needed to do this project.

	Circumference (c) (in feet)	Height (h) (clinometer reading plus your own height)
Tree 1		
Tree 2		
Tree 3		
Tree 4		
Tree 5		
Tree 6		
Tree 7		
Tree 8		
Tree 9		
Tree 10		

Equations:

Diameter: $c = \pi \times d$

$d = \frac{c}{\pi}$ (convert to feet)

Radius: $d = 2 \times r$

$r = \frac{d}{2}$ (in feet)

Area: $A = \pi \times r^2$ (in ft^2)

Volume: $V = A \times h$ (in ft^3)

Letters:

c = circumference
π = pi (3.14)
d = diameter
r = radius
A = area
V = volume
h = height
bd ft = board feet

Number of board feet in a tree trunk $= \frac{12 \text{ bd ft}}{ft^3} \times V$

(There are 12 board feet in one cubic foot of wood.)

Name ______________________ Date ______________________

TREE MEASUREMENT
Student Answer Sheet

You have collected field data on the circumference and height of ten different trees. This information is recorded on the assignment sheet that was used during the first part of the activity. Record the values for *c* and *h* for each tree in the proper spaces below. Then, using the equations provided on the assignment sheet, calculate values for *d, r,* r^2*, A, V,* and *bd ft* for each tree. Record these answers below also. Remember, each number must have units to be correct. Put your calculations on notebook paper and hand them in with this answer sheet.

Tree 1

c = ______________________ r^2 = ______________________

h = ______________________ A = ______________________

d = ______________________ V = ______________________

r = ______________________ bd ft = ______________________

Tree 2

c = ______________________ r^2 = ______________________

h = ______________________ A = ______________________

d = ______________________ V = ______________________

r = ______________________ bd ft = ______________________

Tree 3

c = ______________________ r^2 = ______________________

h = ______________________ A = ______________________

d = ______________________ V = ______________________

r = ______________________ bd ft = ______________________

Tree 4

c = ______________________ r^2 = ______________________

h = ______________________ A = ______________________

d = ______________________ V = ______________________

r = ______________________ bd ft = ______________________

Tree 5

c = ______________________ r^2 = ______________________

h = ______________________ A = ______________________

d = ______________________ V = ______________________

r = ______________________ bd ft = ______________________

Name ______________________ Date ______________________

TREE MEASUREMENT
Student Answer Sheet (continued)

Tree 6

c = ______________ r^2 = ______________

h = ______________ A = ______________

d = ______________ V = ______________

r = ______________ bd ft = ______________

Tree 7

c = ______________ r^2 = ______________

h = ______________ A = ______________

d = ______________ V = ______________

r = ______________ bd ft = ______________

Tree 8

c = ______________ r^2 = ______________

h = ______________ A = ______________

d = ______________ V = ______________

r = ______________ bd ft = ______________

Tree 9

c = ______________ r^2 = ______________

h = ______________ A = ______________

d = ______________ V = ______________

r = ______________ bd ft = ______________

Tree 10

c = ______________ r^2 = ______________

h = ______________ A = ______________

d = ______________ V = ______________

r = ______________ bd ft = ______________

Name ______________________ Date ______________________

MATHEMATICS IN THE PARK
Final Test

1. Your teacher has identified a particular tree that will be the subject of this test.
2. Measure the circumference of the tree, using string and a yardstick or tape measure.
3. Using what you have learned about pacing, pace 100 feet from the base of the tree.
4. Using your homemade clinometer, sight the highest point of usable wood and read the height (h), in feet, directly from the scale. Remember to add your own height to this number.
5. Calculate the *diameter* of the tree at its base by using a related equation.
   $c = \pi \times d$ (This answer should be in *feet.*)
6. Calculate the *area* of a cross section of the tree at its base. (In other words, the area of a circle.)
   $d = 2 \times r$ $\quad A = \pi \times r^2$ (This answer should be in *square feet.*)
7. Calculate the volume of the tree's trunk (in other words, the volume of a cylinder).
   $V = A \times h$ $\quad h =$ the height of the tree (This answer should be in *cubic feet.*)

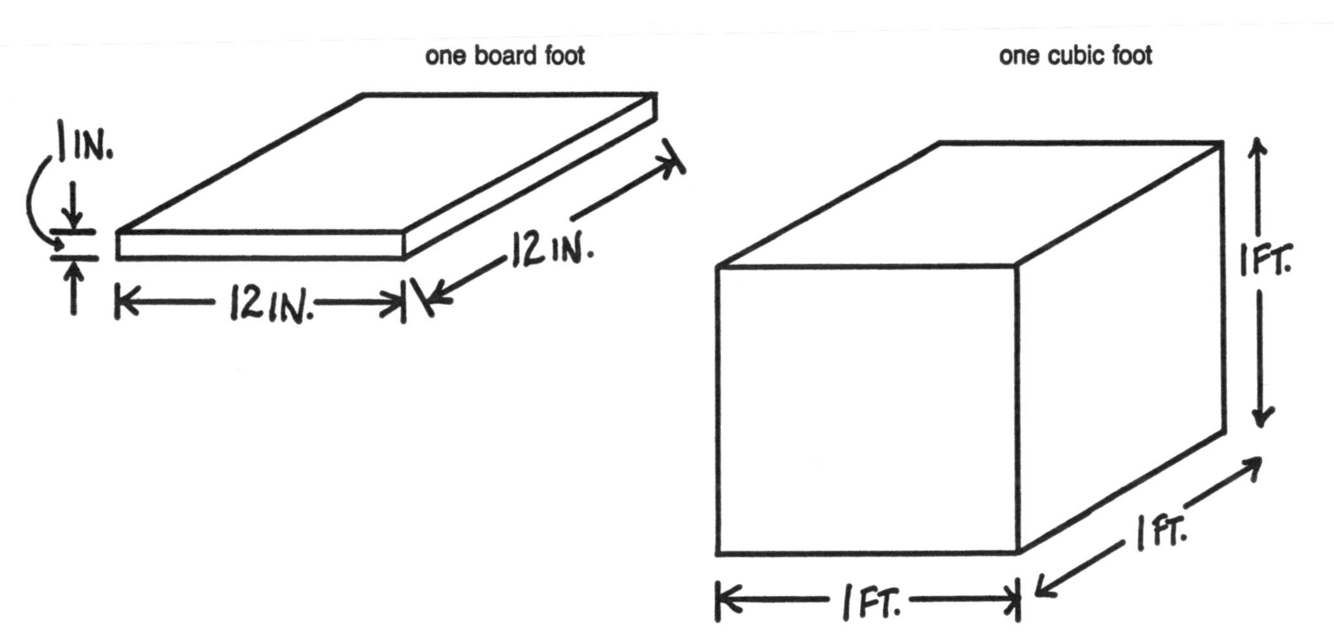

8. One cubic foot equals how many board feet?
9. How many board feet are there in the tree trunk?

Name ______________________ Date ______________________

MATHEMATICS IN THE PARK
Final Test—Student Answer Sheet

Record your answers to the test questions on the spaces provided. Show all of your calculations below each answer.

1. Tree description ______________________

2. c = ______________________

3. Did you pace 100 feet? ______________________

4. h = ______________________

   Are c and h accurately recorded? ______________________

5. d = ______________________

   calculations:

6. A = ______________________

   calculations:

7. V = ______________________

   calculations:

8. One cubic foot = ______________________ board feet.

   calculations:

9. How many board feet are in the tree trunk? ______________________

   calculations:

Name ______________________ Date ______________________

HOW TO BUILD A CLINOMETER

1. Materials:
   a. rectangular piece of stiff cardboard or plywood (approximately 10 inches by 12 inches)
   b. string (12-inch)
   c. soda straw
   d. weight (bolt, nut, washer, sinkers, or the like)
   e. clinometer scale (provided on the following page)
2. Procedure:
   a. Glue or paste the scale to the piece of cardboard.
   b. Fasten the weight to one end of the string.
   c. Fasten the other end of the string to the scale, at the point marked "plumb line attached here." This should be very close to the "top" of the clinometer.
   d. Glue a soda straw along the top of the clinometer.
3. Example:

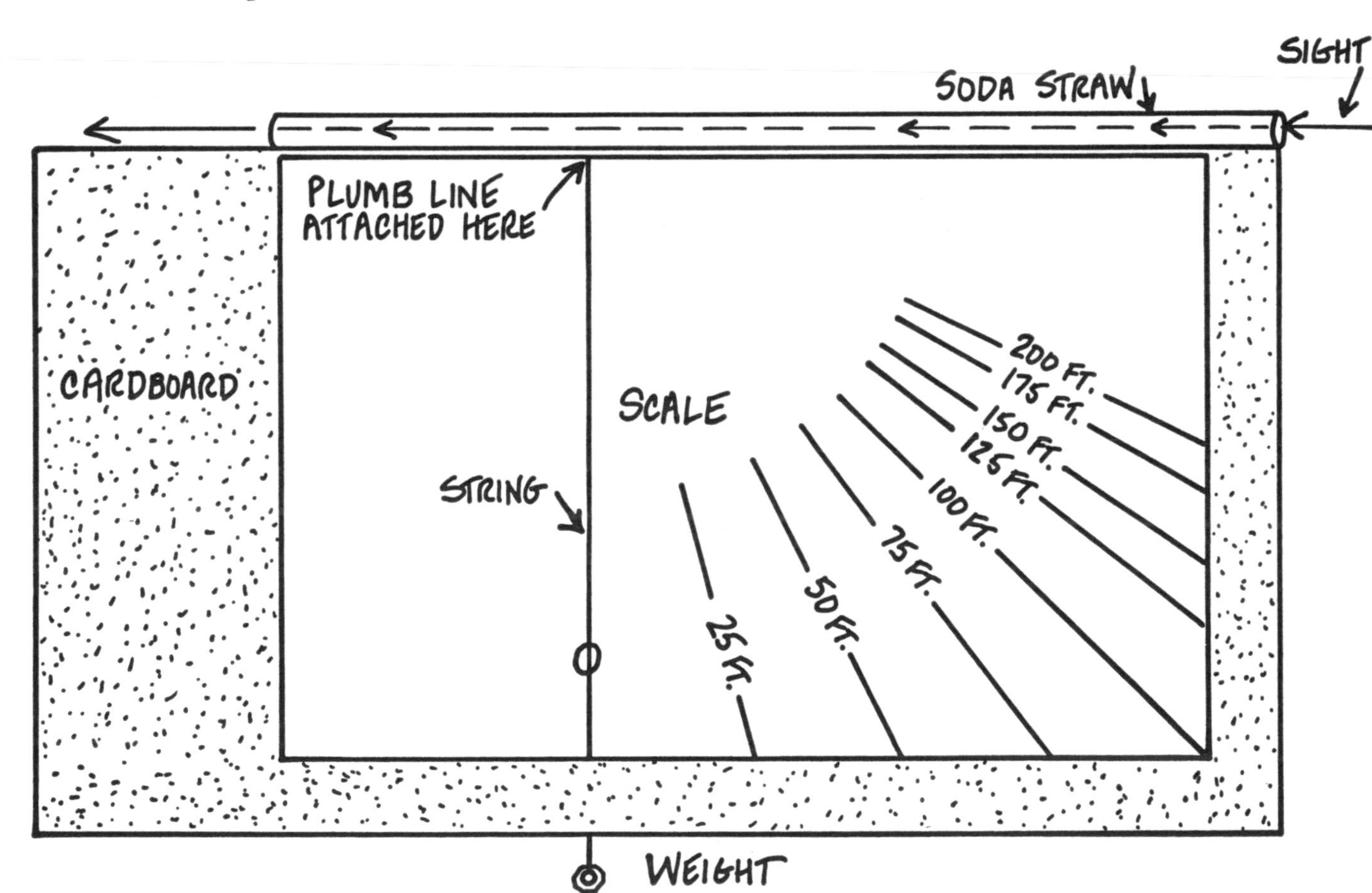

If you cannot stand 100 feet from the base of the tree, use the following equation to find the actual height:

$$\text{ACTUAL HEIGHT} = \text{HEIGHT YOU READ} \times \frac{\text{DISTANCE FROM BASE}}{100 \text{ ft}}$$

plus YOUR HEIGHT

Example: You are standing 75 feet from the tree and get a reading of 85 feet.

ACTUAL HEIGHT (from eye level) $= 85 \text{ ft} \times \frac{75 \text{ ft}}{100 \text{ ft}}$

$= 85 \text{ ft} \times .75$

$= 63.75 \text{ ft}$

This could be rounded to 64 feet.
To find the height of the tree from the ground you must add your height to 64 feet.

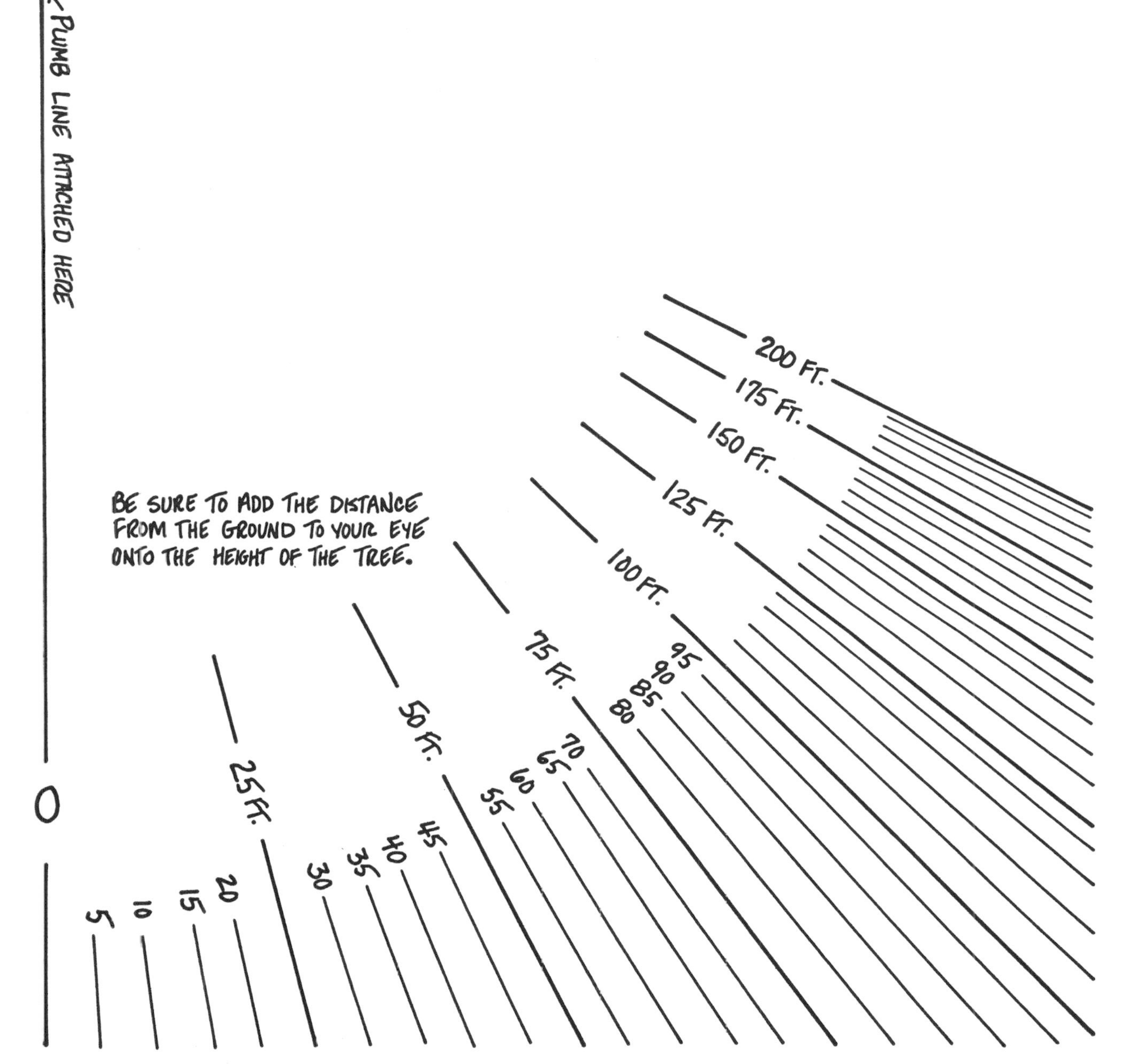

M-9

INDEPENDENT/INDIVIDUALIZED MATHEMATICS PROJECTS

Teacher Preview

Project Topics: Opinion Poll
Creating Story Problems
Mathematics from Current Events

General Explanation:

There are many ways to present these projects, from giving students the handouts and a due date three weeks hence, to making daily lesson plans and doing them in class. They are presented here as independent projects requiring three hours of class time: assignment, mid-project report, and due date. Each of the projects is self-explanatory, but the handouts are brief and make the assumption that the students who are using them have previous experience in working from project outlines. The projects are designed to allow students to decide such things as methods of recording information, the number of information sources, methods of presentation, and types and length of reports.

Length of Each Project: 3 hours

Level of Independence: Advanced

Goals:

1. To allow students to work on their own to produce projects that are based on mathematics.
2. To place emphasis on independent learning skills.
3. To allow students to make some decisions about what they learn.
4. To promote the concept of kids teaching kids.

During This Project Students Will:

1. Plan projects that require the application of mathematics.
2. Assemble information for producing mathematics problems.
3. Organize step-by-step solutions to mathematics problems.
4. Prepare well-organized presentations.

Skills:

Basic mathematics skills
Collecting data
Persistence
Taking care of materials

Interviewing
Library skills
Listening
Observing
Neatness
Organizing
Setting objectives
Selecting topics
Divergent-convergent-evaluative thinking
Following and changing plans
Identifying problems
Meeting deadlines
Working with limited resources
Accepting responsibility
Concentration
Controlling behavior
Following a project outline
Individualized study habits
Time management
Personal motivation
Self-awareness
Sense of "quality"
Setting personal goals
Creative expression
Creating presentation strategies
Drawing/sketching/graphing
Poster making
Public speaking
Self-confidence
Teaching others
Accuracy
Solving for an unknown
Using equations and related sentences
Working with fractions and decimals

Handouts Provided:

- "Opinion Poll Assignment Sheet"
- "Opinion Poll: Topics, Population, and Questions (Example)"
- "Opinion Poll: Data Collection Sheet (Example)"
- "Opinion Poll: Graphs (Examples)"
- "Creating Story Problems Assignment Sheet"
- "Mathematics from Current Events Assignment Sheet"

PROJECT CALENDAR:

HOUR 1: ____________	HOUR 2: ____________	HOUR 3: ____________
Students are given their assignment sheets, the project is discussed, and due dates are set. HANDOUTS PROVIDED	Mid-project report: discussion of how independent projects are going. Help is offered if it is needed. This hour is optional.	Students present and turn in their finished projects. An additional hour may be necessary if presentations are lengthy. STUDENTS TURN IN WORK
HOUR 4: ____________	HOUR 5: ____________	HOUR 6: ____________
HOUR 7: ____________	HOUR 8: ____________	HOUR 9: ____________

Lesson Plans and Notes

HOUR 1: These projects are designed to be done independently, on the students' own time. Spend this hour handing out assignment sheets and explaining the projects. Since there are three projects, students can be given a choice of which to work on. Or, the entire class can work on the same project, reserving the other two for future use. During this hour set the due dates: one for a mid-project report (optional) and one for final projects to be presented and turned in.

Notes:

- Be sure to evaluate the topics that students choose. Don't let anyone pursue a topic that has insufficient information available or that is frivolous in nature.
- It is left to you to develop a grading system for these projects, based on the areas of emphasis that have priority in your classroom.
- These projects are written in the form of "learning contracts"; requiring students' signatures is not necessary, but it is a good idea if work is to be done independently outside the classroom.

HOUR 2: This hour is optional, but it is suggested that it be used, particularly with students who have little independent study experience. Ask students how their projects are going and if any insurmountable roadblocks have been encountered. It is a good idea to have students write brief progress reports to be handed in. This will help determine if further classroom time is necessary.

HOUR 3: Students turn in their finished projects. If presentations have been assigned, they are given during this hour. If the presentations are lengthy, they are likely to require more than one hour to complete.

General Note About This Project:

- Each of these projects has potential as a full-class project. Look all three over carefully for their usefulness in your particular situation. You may want to develop an opinion poll that an entire class can conduct (30 students $\times$ 25 respondents per student = 750 people being polled; a significant number); or, you may decide to have students make a series of story problems to be used in other classrooms or with younger students. The current events project may be used in conjunction with the social studies curriculum to help emphasize specific sets of statistics or numerical facts. There are many ways these projects can be used, so don't be limited to thinking of them only as independent study projects for individual students.

Name ______________________________ Date ______________________________

OPINION POLL
Assignment Sheet

Opinion polls are used extensively in the modern world. Advertisers, politicians, business people, special interest groups, social scientists, news organizations, and government departments all use polls to find out what people think about certain issues. The information that comes from a poll is not very useful, however, until it is put into visual form. Graphs are commonly used to illustrate the results of a poll. They show the differences and similarities between two or more groups of people. Mathematics plays an important part in producing such a graph; raw data must be changed into numbers or percentages before it can be used to illustrate what has been discovered about people's attitudes and opinions.

For this project, choose two topics, make a list of specific questions and interview a number of people (the more the better). After collecting a sufficient amount of data, you will organize it, calculate percentages, produce graphs, and try to draw some conclusions about your findings. Here are the details of the assignment.

Name ______________________________ Date ______________________________

OPINION POLL

Assignment Sheet (continued)

I. Decide on two topics to cover in your poll.

A. They may be about local, national, or world issues and events.

B. They should be about one of the examples below. If you choose different topic areas, have them approved by the teacher.

1. The nuclear weapons race/defense
2. Prayer in public schools
3. Election races/politicians
4. The economy
5. The president
6. Students'/children's rights
7. "Back to basics" in school
8. Television/music
9. Attitudes about foreign countries
10. Taxes/social security/federal deficit
11. Environmental issues
12. Capital punishment
13. Crime prevention
14. Issues in education
15. Professional sports

II. Choose a polling population.

A. Break the population into two groups that will make for interesting comparisons. (More than two groups makes this project quite complicated.)

B. Question at least 12 people from each group to make the poll meaningful. It would be better to interview 25 or more people from each group. See the following list for suggestions of group categories.

1. Girls under 13
2. Girls 13 and over
3. Boys under 13
4. Boys 13 and over
5. Adult women
6. Adult men
7. Married people
8. Unmarried people
9. College students
10. Students in your school
11. Students from other schools
12. Parents
13. Blue-collar workers
14. Ethnic groups (Blacks, Hispanics, Italians, Orientals, Indians, etc.)
15. Any other group of people that you would like to include

III. Write questions.

A. Make the questions short and to-the-point.

B. Word each question carefully so the answer can be specific.

C. You need not write very many questions; four (two for each topic) is enough. The secret of a useful poll is to survey a lot of people with a few well-written questions.

Name ______________________ Date ______________________

OPINION POLL

Assignment Sheet (continued)

D. Tell the people being questioned what the response choices are. Questions should be worded so that these responses make sense. Here are four types of responses that might be used:
   1. Yes or no; agree or disagree
   2. On a scale from 0–10
   3. Choice of four or five possible answers, like: "poor–fair–good–very good–excellent," or "strongly agree–agree–no opinion–disagree–strongly disagree"
   4. Multiple choice

E. Some "questions" can actually be statements that people are asked to agree or disagree with. For example: "On a scale from 1 to 10, with 1 being total disagreement and 10 being total agreement, rate your feelings about this statement: The United States should spend whatever is necessary to move ahead of the Soviet Union in the nuclear arms race."

F. Questions should be reviewed by your teacher before beginning the poll.

IV. Conduct the poll and compile data. Design a data collection sheet on which to record responses. An example is provided for you on a separate handout. Be very careful, because precise calculations, meaningful comparisons, and useful graphs can come only from accurate information.

V. Graph the results. Examples are provided for you to examine before designing your own.

A. Be creative and try to find as many different ways as possible to show the results of your work.

B. Calculate percentages that show how each group reacted to each question. For example, what percent of adult women thought that schools should place much more emphasis on basic skills? What percent of girls under 13 felt this same way?

C. When your work is finished, make a classroom display of it for others to see. You may also be asked to make an oral presentation.

IMPORTANT NOTE: Be sure to work closely with your teacher on this project to ensure its success. Don't be afraid or reluctant to ask for help if you run into problems.

______________________ has completed all assignments, is passing all subjects, continues to do satisfactory work, and is allowed to begin an independent project titled "Opinion Poll."

Student ______________________

Teacher ______________________

Due Date ______________________

OPINION POLL: TOPICS, POPULATION, AND QUESTIONS
Example

by ______Mary Doe______

TOPIC I–Television

TOPIC II–Environmental Issues

GROUPS TO BE POLLED (Population)

1. Girls 13 and over
2. Boys 13 and over

Topic I, Question 1

In your opinion, does the amount of television children watch expose them to too much violence?

YES NO

Topic I, Question 2, Part A

Which one of the following types of programming do you enjoy watching on television the *most*?

A. Sports	E. T.V. movies
B. Soap operas	F. News
C. Rock videos	G. Sitcoms
D. Cartoons	H. Drama

Topic I, Question 2; Part B

Which type of programming do you enjoy watching on television the *least*?

A. Sports	E. T.V. movies
B. Soap operas	F. News
C. Rock videos	G. Sitcoms
D. Cartoons	H. Drama

Topic II, Question 1

On a scale of 1 to 10, how serious do you believe the "acid rain" problem is? (1 is *not serious at all;* 10 is *extremely serious.)*

1 2 3 4 5 6 7 8 9 10

Topic II, Question 2

How would you rate the government's efforts to improve the quality of our environment?

Poor Fair Good Very Good Excellent

OPINION POLL: DATA COLLECTION SHEET
Example

By Mary Doe

I am conducting an opinion poll for a class project. You can help me by honestly answering two questions about television and two questions about environmental issues. I will ask the questions and record your answers on my Data Collection Sheet. Your name will *not* be used in any way.

Questions	Answer Choices	Girls 13 or Older	(T)	Boys 13 or Older	(T)
TELEVISION QUESTION 1: In your opinion, does the amount of television children watch expose them to too much violence?	YES	𝍸 𝍸 𝍸 |||	18	𝍸 𝍸 ||	12
	NO	𝍸 |||	8	𝍸 𝍸 𝍸		16
TELEVISION QUESTION 2: Part A Which one of the following types of programming do you enjoy watching on television the *most*?	SPORTS		0	𝍸 𝍸 ||	12
	SOAP OPERAS	𝍸 𝍸 |||	13		0
	ROCK VIDEOS	𝍸 ||	7	𝍸 𝍸	10
	CARTOONS	||	2	||	2
	T.V. MOVIES	|||	3			1
	NEWS			1	||	2
	SITCOMS		0			1
	DRAMA		0		0
TELEVISION QUESTION 2: Part B Which one of the following types of programming do you enjoy watching on television the *least*?	SPORTS	𝍸 𝍸	10		0
	SOAP OPERAS		0	𝍸		6
	ROCK VIDEOS		0		0
	CARTOONS		0		0
	T.V. MOVIES		0		0
	NEWS	𝍸 𝍸 ||||	14	𝍸 𝍸 𝍸 𝍸	20
	SITCOMS		0		0
	DRAMA	||	2	||	2

OPINION POLL: DATA COLLECTION SHEET
Example (continued)

By Mary Doe

Questions	Answer Choices	Girls 13 or Older	(T)	Boys 13 or Older	(T)
ENVIRONMENTAL ISSUES QUESTION 1: On a scale of 1 to 10, how serious do you believe the "acid rain" problem is? (1 is not serious at all; 10 is extremely serious.)	1	𝍸 I	6	𝍸 II	7
	2		0		0
	3	III	3	IIII	4
	4		0	II	2
	5	II	2	I	1
	6	IIII	4	𝍸	5
	7	II	2	I	1
	8	IIII	4	III	3
	9		0	I	1
	10	𝍸	5	IIII	4
ENVIRONMENTAL ISSUES QUESTION 2: How would you rate the government's efforts to improve the quality of our environment?	POOR	IIII	4	III	3
	FAIR	𝍸 I	6	𝍸	5
	GOOD	𝍸 𝍸	10	𝍸 𝍸 II	12
	VERY GOOD	III	3	𝍸	5
	EXCELLENT	III	3	III	3

Total number of girls 13 or older interviewed: 26

Total number of boys 13 or older interviewed: 28

OPINION POLL: GRAPHS
Examples

by Mary Doe

Responses from 26 girls and 28 boys, 13 or older

TOPIC: Television

Date ____________

TELEVISION: QUESTION 1

In your opinion, does the amount of television children watch expose them to too much violence?

GIRLS' RESPONSE: Pie Graph

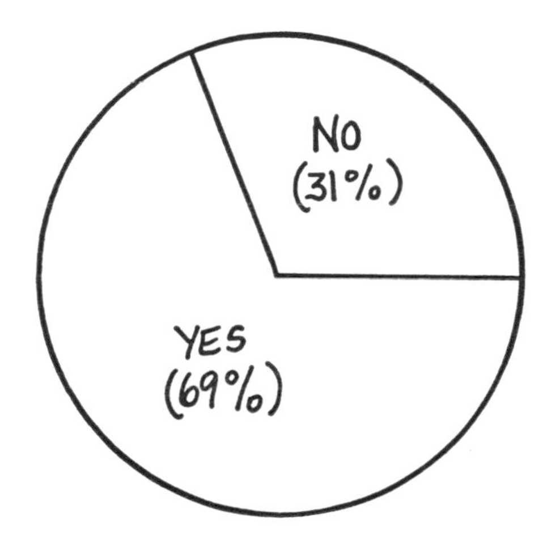

TELEVISION: QUESTION 1

In your opinion, does the amount of television children watch expose them to too much violence?

BOYS' RESPONSE: Pie Graph

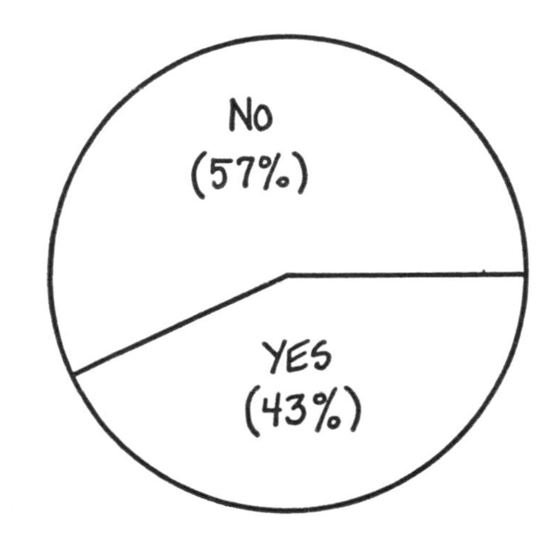

TELEVISION: QUESTION 2, Parts A and B

Which type of T.V. programs do you *most* like to watch, and *least* like to watch?

GIRLS' RESPONSE: Vertical Bar Graph

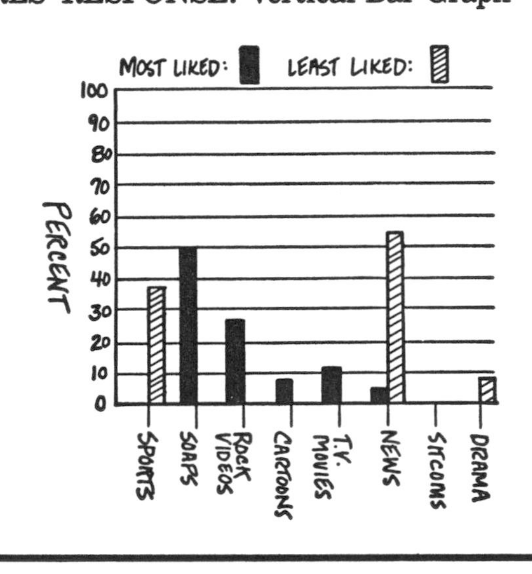

TELEVISION: QUESTION 2, Parts A and B

Which type of T.V. programs do you *most* like to watch, and *least* like to watch?

BOYS' RESPONSE: Vertical Bar Graph

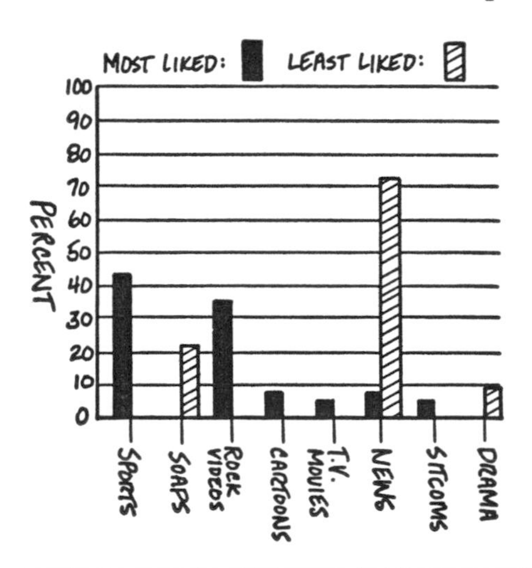

STUDENT OPINION POLL: GRAPHS
Examples (continued)

by Mary Doe

Responses from 26 girls and 28 boys, 13 or older

TOPIC: Environmental Issues

ENVIRONMENTAL ISSUES:
QUESTION 1

On a scale of 1 to 10, how serious do you believe the "acid rain" problem is?

GIRLS' RESPONSE: Horizontal Bar Graph

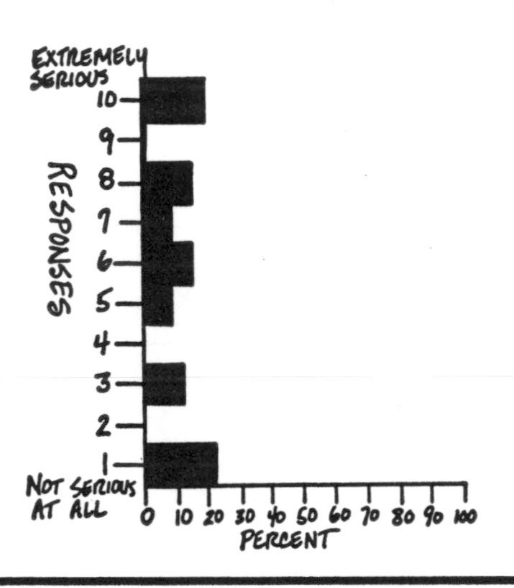

ENVIRONMENTAL ISSUES:
QUESTION 1

On a scale of 1 to 10, how serious do you believe the "acid rain" problem is?

BOYS' RESPONSE: Horizontal Bar Graph

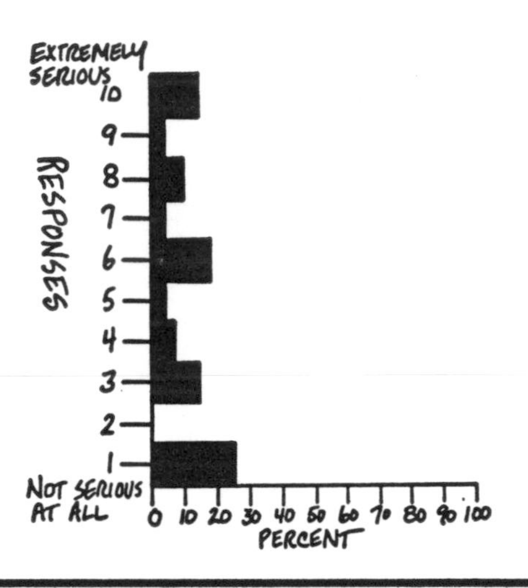

ENVIRONMENTAL ISSUES:
QUESTION 2

How would you rate the government's efforts to improve the quality of our environment?

GIRLS' RESPONSE: Pie Graph

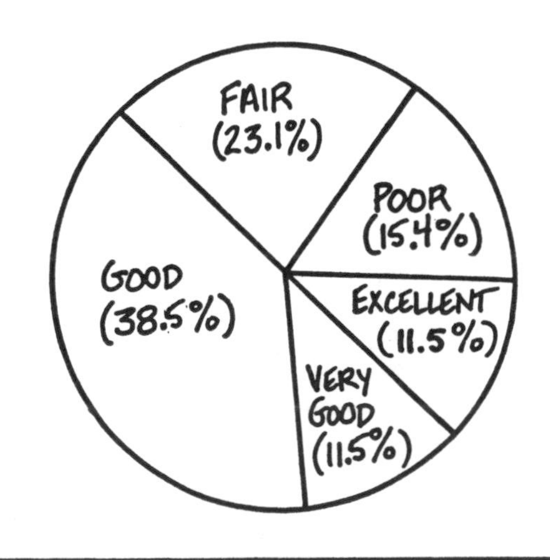

ENVIRONMENTAL ISSUES:
QUESTION 2

How would you rate the government's efforts to improve the quality of our environment?

BOYS' RESPONSE: Pie Graph

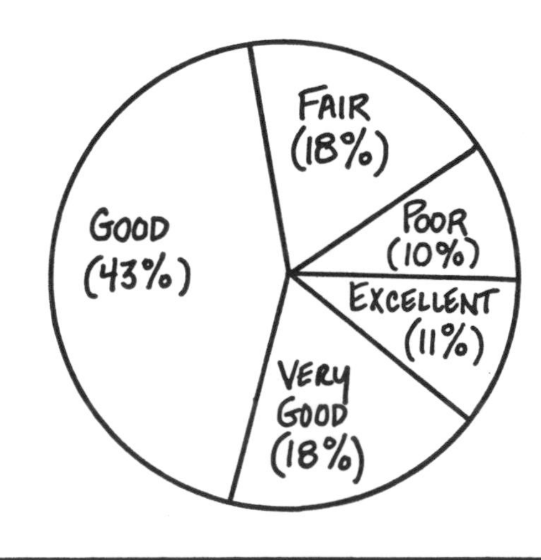

Name ______________________________ Date ______________________________

CREATING STORY PROBLEMS
Assignment Sheet

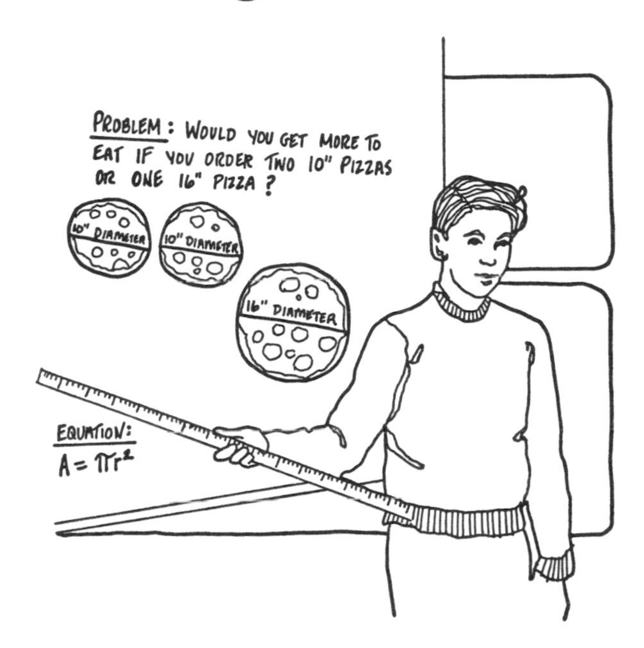

Nearly every mathematics problem in the real world has a story behind it. The reason the problem needs to be solved is because the answer is necessary to complete a project or proceed with an activity. People are constantly using mathematics to determine if an undertaking is possible, or to verify that something has been done accurately. Examples are endless: Given current interest rates, can we afford a new car? How big should the rooms of a new house be? Is electricity the most efficient kind of energy? How fast is America's population increasing? Will a certain chemical reaction be explosive? Of course, the list goes on and on.

Everyone has been faced with mathematics textbooks that offer seemingly endless pages of problems to solve. These problems have no story and so they appear lifeless and without purpose (other than to teach problem-solving skills!). But when faced with a *real* problem, the value of mathematics skills becomes immediately apparent. It is at these times that the ability to solve story problems is needed; it is then that mathematics becomes both useful and fun.

For this project you will create six challenging story problems that can be solved by students your age. Here is the assignment.

I. Identify two subject areas that the problems will cover. You will create three story problems for each area. See the following list for suggestions.

A. Land surveying
B. Automobiles
C. Carpentry
D. Economics
E. Sports
F. Chemistry
G. Physics
H. Weather
I. Astronomy
J. History
K. Electricity
L. Current events

Name ______________________________ Date ______________________

CREATING STORY PROBLEMS

Assignment Sheet (continued)

M. Social data: population, unemployment, public opinion, statistics, education, income, and so forth
N. Politics
O. Geography
P. Basic "living" problems: family budget and income; costs of goods and services
Q. Anything else you can think of

II. Decide what kinds of mathematics problems to emphasize:

A. Addition, subtraction, multiplication, division
B. Fractions
C. Decimals
D. Equations
E. Related sentences
F. Geometry
G. Higher levels of mathematics

III. Collect data for your problems and make a bibliography that records where the information came from. If you want to find out how long it will take to fly at the speed of light from Earth to Pluto, several facts are needed: How fast is the speed of light? How far away is Pluto? Record the facts on notecards for easy access.

IV. Write six problems and supply each with a step-by-step solution. Create three problems for each of the two subject areas you have chosen.

A. Record the equations that apply to the problems.
B. Record all of the information that is to be supplied for the person solving the problems.
C. Make drawings to help illustrate each problem.
D. Solve each problem to your own satisfaction, and show the solution, step-by-step, on paper.

V. Prepare a presentation to the class. For example:

A. Present each problem, one at a time.
B. Ask students to solve the problems.
C. Ask volunteers for their answers.
D. Demonstrate how each problem should be solved, using diagrams and showing calculations on the board.

______________________________ has completed all assignments, is passing all subjects, continues to do satisfactory work, and is allowed to begin an independent project titled "Creating Story Problems."

Student ______________________________

Teacher ______________________________

Due Date ______________________________

Name ______________________ Date ______________________

MATHEMATICS FROM CURRENT EVENTS
Assignment Sheet

Current events is a brief term for describing everything that is happening right now in the modern world. As everyone knows, much is happening. Perhaps less well known, however, is how much mathematics is involved in the current events of the world. We live in a technological, computerized age that relies heavily on numbers. These numbers supply nearly endless opportunities for making comparisons and calculations. Studying current events is an interesting undertaking, and applying mathematical problem solving to such a study can yield fascinating insights into some of the things that are going on around us.

This is a mathematics research project. Choose a topic and find enough data about that topic to create at least five story problems (complete with solutions). Also, record where your facts came from (in other words, provide a bibliography). Here are the details of the assignment.

I. Choose a current events topic. Some examples are

A. The Middle East
B. The economy
C. The nuclear arms race
D. National elections
E. Environmental pollution
F. Population
G. The space program
H. Energy consumption
I. Imports and exports
J. The national economy

Name ______________________ Date ______________________

MATHEMATICS FROM CURRENT EVENTS

Assignment Sheet (continued)

II. Study newspaper and magazine articles about the topic and record facts that might lead to a good mathematics problem. Put the facts on notecards. For example, if your topic is the Middle East, you might find these facts:

A. The population of Israel
B. The land area of Syria
C. The number of Palestinians in Lebanon
D. The amount of exports Egypt sends out each year
E. The number of Muslims there are in the Middle East, or in the world
F. The amount of oil produced per year in Saudi Arabia

III. Think of ways these numbers can be used with other information that can be found in almanacs, encyclopedias, and other reference materials. For example:

A. How much larger is America's population than Israel's?
B. What is the average population per square mile in Syria?
C. What percentage of *all* Palestinians does the number living in Lebanon represent?
D. How much of Egypt's economy is based on exports?
E. Are there more Christians or Muslims in the world? What is the proportion of Christians to Muslims?
F. What percentage of the world's yearly oil production comes from Saudi Arabia?

IV. Write five mathematics problems about the topic you have chosen.

A. Record the equations that apply to the problems.
B. Record all of the information that is to be supplied for someone solving the problems.
C. Make drawings to help illustrate each problem.
D. Solve the problems and show solutions to each with carefully arranged step-by-step procedures. Hand in the completed problems with answers.

V. Prepare a presentation to the class. For example:

A. Present each problem, one at a time.
B. Ask students to solve the problems.
C. Ask volunteers for their answers.
D. Demonstrate how each problem should be solved, using diagrams and showing calculations on the board.

______________________ has completed all assignments, is passing all subjects, continues to do satisfactory work, and is allowed to begin an independent project titled "Mathematics from Current Events."

Student ______________________

Teacher ______________________

Due Date ______________________

M-10

ADDITIONAL IDEAS FOR MATHEMATICS PROJECTS

The need to conserve space prevents the following ideas from being presented as fully written projects. However, teachers who are willing to spend time to produce student assignment sheets can use many of these ideas for independent study projects. Those who don't mind the extra work of developing lesson plans can build some of the ideas into the curriculum. Most of the project ideas are based upon material already presented in this book; they have many applications and are offered as suggestions to be used in whatever way is most beneficial to your mathematics program.

FRACTIONS AND DECIMALS: A CLASSROOM GAME

I. Have students produce materials that can be used with other students. Use correction fluid to erase the dimensions on each handout and make copies for students. Their instructions are to

- A. Create a game for younger students (like the one in this book but with simpler numbers).
- B. Or, create a new game for classmates.
- C. Or, create a complex game for accelerated or selected students.

II. Replace the dimensions on the handouts with numbers that are in line with the skill development of the students in your class.

III. Draw each figure on a rectangular coordinate system and have students locate points (corners of closed figures) and calculate slopes and distances. From this information perimeters and areas can be calculated.

SIMPLE EQUATIONS

I. Conduct a research project:

- A. Assign each student one equation.
- B. Explain what the equation represents and what field of science or mathematics it comes from.
- C. Require students to conduct research and find out everything possible about the equation.
- D. Set a due date when students report to the class about their equations.

II. Conduct experiments or investigations to collect actual data to use in certain equations. You will be amazed how many of the equations used in this project lend themselves to this application. As one brief example: P.E. = mgh can be demonstrated by holding a book above the floor. Data can be collected with a scale and a yardstick. The acceleration of gravity is constant: 32 feet per second per second (32 ft/sec/sec).

MATHEMATICS TREASURE HUNT

I. Reconstruct the treasure map, problem sets, and clues to incorporate a higher level of mathematics. The game was originally designed to teach basic trigonometry: the line segments were all parts of right triangles and the clues supplied angles and lengths of other legs. A map could also be developed over a rectangular coordinate system with obvious possibilities for distances between points, midpoints of line segments, slopes, angles between two lines, and so forth.

II. Have students create their own game by developing a number of problem sets, clues, and a map.

SECTIONS AND ACRES

I. Have students create an imaginary continent or an imaginary country. Together with the class establish a set of principal meridians and base lines, and then show further subdivisions through a series of scale drawings. This project can be combined with a geography unit, a social studies course (by inventing groups of people for the country), or a science course (by populating the country with plants and animals), or by describing landforms, weather, geologic activity, and so forth.

II. Take students outside to measure parcels of land, to calculate how many acres are in them, or what fraction of a section they represent.

III. From local realtors or bankers, find out what land in various areas of your community is selling for. Incorporate this information into section and acre problems.

PROTRACTOR–COMPASS–RULER

I. Have a poster-making contest. The rules require that all lettering be neatly arranged on parallel lines, and that the poster display consist of drawings made with a protractor, a compass, and a ruler.

II. Have students create their own point location problems. They can be as complex as the students wish to make them.

III. Introduce students to a rectangular coordinate system (on an X-Y axis) by teaching them about number pairs and positive and negative numbers. A game called four-across can be played on graph paper divided into quadrants:

it is played like tic-tac-toe, only it takes four in a row to win. Before making a mark, a student writes the number pair for a particular point on the graph paper; that is, (5,2) or (-6,3). This is another form of point location. The same game can be played using degrees of longitude and latitude (6°N, 12°W), which fits in nicely with the Sections and Acres project.

TREE MAPPING

I. Make a map of the school grounds or the neighborhood, by locating the corners of buildings, sidewalks, power poles, fire hydrants, manhole covers, and so forth.

II. Draw a base line through a portion of a detailed topographic map, and have students make a large-scale drawing of a specified section of it. This is done by lightly drawing lines from a randomly selected point on the base line (point "A") to various geographical points on the map, and then measuring and recording angles and distances. This information can be used to redraw the section of the map on a larger scale. Excellent topographic maps may be obtained from the U.S. Geologic Survey, Department of the Interior, Washington, D.C., 20242. Send first for a catalog.

III. Carrying this idea a bit further, students can construct a model of a geographic area, based on their scale drawings and the elevations and details provided on the topographic maps. This makes an excellent project for students who can work independently; they can build models (out of cardboard, papier-mâché, plaster of Paris, and other materials) of virtually any spot in the United States.

LAND SURVEYING

I. Place wooden stakes in the ground (in any geometric shape) prior to the beginning of the project. Have students locate the points, take field notes, and make scale drawings or maps from their notes.

II. Give students the "Teacher Answer Sheets" that are provided and have them conduct "surveys" with protractors and rulers. Students' field notes are turned in, which should follow closely the instructions on the "Student Assignment Sheet."

III. Let students devise their own surveying problems, and perhaps present their problems to another class. Or, after conducting Land Surveying with your students, it could be arranged for *them* to instruct younger students on the use of a transit and the concept of locating points with angles and distances.

IV. Set up an actual land survey, beginning from known property corners: measure angles and distances and locate buildings, then make a scale drawing.

MATHEMATICS IN THE PARK (TREE MEASUREMENT)

I. The height of any object can be estimated with a clinometer, so the volume of buildings or other objects can also be calculated.

II. Combine this project with Tree Mapping and a tree identification program to produce the nucleus for a forestry unit that is highly focused on mathematics.

III. The clinometer can be used to introduce students to some basic concepts of trigonometry: proportions, similar triangles, right triangles, and trigonometric functions. Trigonometry can be used to calculate the height (see "a" in the figure below) of an object by measuring the distance (b) from its base and an angle (A) to its highest point. The tangent of that angle, times the distance from the base, equals the height:

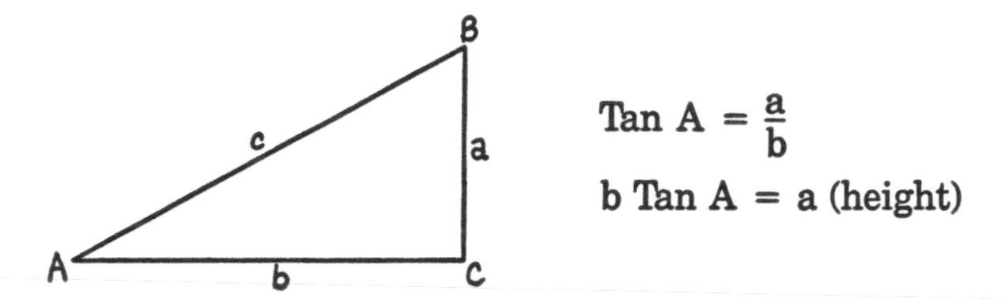

The vertical angle can be measured with a simple clinometer that is built just like the one for Mathematics in the Park, only with degrees marked on the scale instead of feet. When using the trigonometry method of calculating height there is no need to be a certain distance from the base of the object being measured.

OPEN HOUSE OR SPECIAL PRESENTATION

This final project idea is based upon the theory that it is important for students to display, exhibit, or present their work to others. This is accomplished to some extent with classroom and school activities, but at some point it is also important to let students show an adult population what they have done. Usually this type of open house is reserved for subjects that are easily put into visual form. That is why science fairs are popular and why social studies classes have their numerous programs based on the foods, clothing, and customs of other cultures, political debate, role playing, world geography, and many other topics.

Mathematics, too, can be presented visually, and a mathematics open house is a worthwhile undertaking. Students can teach mini-lessons; display their posters; demonstrate the use of equipment; report on applications of mathematics in the real world; present the solutions to problems they have created; explain how certain mathematical principles work; and exhibit maps, drawings, graphs, diagrams, equations, and models. You will find that adding a culminating activity like an open house to the curriculum will provide motivation while giving the more accelerated students something to shoot for. This motivational factor should coax students to go beyond the minimum requirements of a project or an assignment, and it will probably encourage some to strike out on their own with independent projects. After all is said and done, it is enthusiasm for learning that lights the flame of creativity and productivity. Interesting, exciting projects, combined with opportunities to show others the work that has been done, produce enthusiasm in many students who might otherwise be uninspired in mathematics. Since it is such an important field of study, it is worth the effort to make it challenging and rewarding.

Appendix

SKILLS CHART: MATHEMATICS

Prerequisite Skills — Students must have command of these skills.

X *Primary Skills* — Students will learn to use these skills; they are necessary to the project.

0 *Secondary Skills* — These skills may play an important role in certain cases.

* *Optional Skills* — These skills may be emphasized but are not required.

	RESEARCH					WRITING & PLANNING			PROBLEM SOLVING											
	COLLECTING DATA	INTERVIEWING	LIBRARY SKILLS	LISTENING	OBSERVING	NEATNESS AND ORGANIZING	SETTING OBJECTIVES	SELECTING TOPICS	ACCURACY	BASIC MATHEMATICS SKILLS	DIVERGE-CONVERGE-EVALUATE	DRAWING STRAIGHT & PARALLEL LINES	FACTORING	FOLLOWING & CHANGING PLANS	IDENTIFYING PROBLEMS	LINEAR & ANGULAR MEASUREMENT	MEETING DEADLINES	MULTIPLICATION & ADDITION PROPERTIES	POINT LOCATION	SCALE DRAWING/MAPPING

SKILLS CHART: MATHEMATICS

PROBLEM SOLVING								SELF-DISCIPLINE										SELF-EVALUATION				PRESENTATION						
SOLVING FOR AN UNKNOWN	USING A COMPASS	USING A HOMEMADE TRANSIT	USING A PROTRACTOR	USING A RULER/STRAIGHTEDGE	USING EQUATIONS AND RELATED SENTENCES	WORKING WITH FRACTIONS AND DECIMALS	WORKING WITH LIMITED RESOURCES	ACCEPTING RESPONSIBILITY	CONCENTRATION	CONTROLLING BEHAVIOR	FOLLOWING PROJECT OUTLINES	INDIVIDUALIZED STUDY HABITS	PERSISTENCE	SHARING SPACE	TAKING CARE OF MATERIALS	TIME MANAGEMENT	WORKING IN GROUPS	PERSONAL MOTIVATION	SELF-AWARENESS	SENSE OF "QUALITY"	SETTING PERSONAL GOALS	CREATIVE EXPRESSION	CREATING PRESENTATION STRATEGIES	DRAWING/SKETCHING/GRAPHING	POSTER MAKING	PUBLIC SPEAKING	SELF-CONFIDENCE	TEACHING OTHERS